IEE MANAGEMENT OF TECHNOLOGY SERIES 17

Series Editors: G. A. Montgomerie
B. C. Twiss

HOW TO COMMUNICATE IN BUSINESS

A handbook for engineers

Other volumes in this series:

HOW TO COMMUNICATE IN BUSINESS

A handbook for engineers

David Silk

The Institution of Electrical Engineers

Published by: The Institution of Electrical Engineers, London,
United Kingdom

© 1995: The Institution of Electrical Engineers

The Institution of Electrical Engineers,
Michael Faraday House,
Six Hills Way, Stevenage,
Herts. SG1 2AY, United Kingdom

British Library Cataloguing in Publication Data

A CIP catalogue record for this book
is available from the British Library

ISBN 0 85296 872 8

Printed in England by Short Run Press Ltd., Exeter

Contents

This book is dedicated to the memory of

David Ellis-Jones

(1919–1992)

A master in the art of communication

Preface

Communication enables people to work together. It underpins the business enterprise, both internally and externally. It is becoming even more important, as we adopt the management styles of quality, team working and empowerment of individuals. Effective communication is not easy, but it is a skill you can develop.

This book aims to help engineers to communicate effectively with nonengineers, in the business context. It will also be useful to people from other technical or numerate backgrounds. It will help you to consider the various methods of communication used in business, and consciously develop your own skills. The approach is practical, although built upon some simple ideas of what we mean by communication, and its role in the business enterprise.

Major chapters deal with spoken communication, written communication, business presentations and business meetings. They cover the main communication skills that an engineer is likely to need, both in a professional capacity and in a management or administrative capacity. Cross-cultural and international aspects are considered. A chapter on the use of information technology deals with IT as a personal aid, and also internal and external networks which operate in the text, data, voice and image modes. Each requires good technique to release its full benefit. All chapters include practical advice and questions to help your self assessment and development. The final chapter suggests how you can construct a personal action plan, and help in the development of your colleagues.

I have tried to make the book practical, and crisply written. I hope it helps to develop your communication skills to meet the challenging needs of modern business.

David J Silk
April 1995

Acknowledgments

Franklin P Jones defined originality as the art of concealing your sources. I do not aspire to that, but there is a practical difficulty: each of us is indebted to everyone with whom we have ever communicated. It is therefore impossible to mention them all.

However, I would like to thank the authors of the listed sources, and those colleagues and friends with whom I have shared communication during eleven years working in management education. More specifically, I would like to thank Mrs Linda Archer and Mrs Nesta Ellis-Jones for their kind permission to use the work of their late husbands, in Figure 3-6 and Section 8.2 respectively. And my family, for their patience.

Chapter 1

Introduction

'Man has three kinds of language: letters, numbers and music'
Peter Calvocoressi (1912-)

1.1 The problem

Often people misunderstand each other. When this happens we talk about a failure of communication. It can happen in a personal relationship, in a wider social context or in the context of our work. In each of these areas the result may be trivial, and even amusing. However, it might equally cause lasting damage.

This book concentrates on the process of communication in business. It analyses why we need to communicate, and what can go wrong. It is aimed at engineers and those from a similar technical or numerate background, because their needs and strengths are different from those of other professional groups. The book will help you to understand the process of communication, and become more effective at it. It is about developing your personal skill in a number of areas.

This introductory chapter is in three sections. The first looks at the general problem of communication in business. The second looks at the challenge this presents to engineers in particular. The third section describes the structure of the rest of the book, and how to use it.

First, the problem. The difficulty often lies in a misunderstanding between two groups of people with different backgrounds and roles:

- Engineers, with a technical background and current responsibilities that may still be mainly in that area
- Nonengineers, with a nontechnical background or with a different type of

professional training, who have responsibilities in nontechnical areas of the business

That's a rather gross simplification, but it does highlight the aspect that I focus on. You would not have read this far unless you were aware of a broad problem of this type. So stop and consider for a moment the barriers that can hinder effective communication between these two groups: engineers and nonengineers. Pause and make a mental list, or better still get a piece of paper and jot your ideas down.

Don't read on until you have given that serious thought! As an engineer you know that understanding a problem is halfway to solving it. Here are some of the barriers that may have occurred to you:

— Engineers like to be precise, and confident of their facts. They are uncomfortable dealing with the 'softer' aspects of management like team building, morale, and the uncertainty of the business environment.
— Related to that, engineers are numerate. They are used to handling numbers, graphs and statistics. Nonengineers are often not very good at this, but are reluctant to admit they don't understand what is being presented to them.
— Nonengineers (especially managers) discuss topics like market growth, business turbulence, poor cash flow, brand awareness and customer loyalty without making it clear what they mean. Even when they define what they mean they often can't quantify it. Engineers find this frustrating.
— Specialists from other disciplines use jargon which obscures what they are saying. They seem unable to communicate with the rest of us in language we can understand, and which relates to the needs of the business. Of course, engineers are never guilty of this themselves! They always express their ideas clearly, in terms which the nonspecialists who are involved can understand, don't they?

These statements are all generalisations. However, you can probably think of incidents from your own experience which support them. The problem is that the business enterprise does not consist of any single specialisation or function. It is a complex system, although it exists for one purpose: to satisfy customers and stakeholders. Of course there are many types of business, but that purpose is essentially the same:

● In a commercial company, if you satisfy your customers they will continue to buy your product (whether that product is a tangible good, or a service). You can then make a profit and satisfy your main stakeholders. You might view the stakeholders as first the owners or shareholders of the business, and secondly the community and environment within which the business operates.
● In a nonprofitseeking enterprise like a charity the emphasis may be even more directly on the customer. But even there if you fail to convince stakeholders such as donors to the charity that you are doing a worthwhile and effective job then you will eventually be put out of business.
● In a public-sector enterprise, the customers are the users of the service which you provide. Usually they are also taxpayers and voters, and thus stakeholders.

Together with other such stakeholders they can eventually put you out of business if they do not like what you do.

Thus a business enterprise must satisfy both customers and stakeholders. If asked which is most important, we must say the customer. It is the customer who has the most immediate control over the success of your business. If the customer is dissatisfied and stops buying your product then your business will fail very quickly. The only exception to this rule is the monopoly supplier, and there aren't many of those left!

A business enterprise must produce and deliver products which satisfy the needs of customers, enabling it to satisfy other stakeholders as well. This is a complex operation. It requires a variety of roles, each needing specialist skills. These include: strategic planning, human resource management, finance and accounting, engineering, production, logistics, marketing, sales and customer support. The specialists and professionals in each of these areas will have a different role to play and a different perspective of the business as a whole. Their attitude to each situation is therefore likely to be slightly different.

These differences can give rise to the communication difficulties mentioned earlier. Here are four examples, corresponding to the four barriers we identified:

Personnel manager: 'I'm concerned about morale. Since we introduced the new organisation structure our people don't seem as committed as they should be. We need to get teams of people working together more effectively. Perhaps you could introduce some measures in your project team, to get them really performing?' Engineer: 'Well, it's difficult when we are so busy just trying to keep things going to plan. Morale seems all right to me; we usually get things done in the end, you know.' Engineer (thinks): 'What on earth does he mean? I can't measure morale, and in any case what causes good or bad morale? If I don't know that, and I can't measure the result, how can they expect me to do anything about it?'

Engineer: 'I've worked out the total cost of producing our widgets. It comes to 2.83 pence each, with a standard deviation of 0.03 pence. Oh, and that assumes that electricity and labour prices are held within 2.5%' General manager: 'I see. That's useful, but I only know to an accuracy of 15% the cost of packaging and delivering the widgets to our customers'. General manager (thinks): 'Why do engineers always put everything in numbers to two decimal places? They don't realise that other aspects of what we do can be measured only in the vaguest way. Anyway, our competitor is now selling widgets for only two pence each. That's my real headache. I bet the engineer doesn't know that.'

Marketing manager: 'I know the name doesn't really reflect what the product now does, but it's a well known brand. Brand awareness is really very important, you know. In fact we now put a brand valuation in our annual company accounts. It can amount to millions. We can't throw all that customer goodwill away.' Engineer: 'But that's silly. Any sane customer will buy the product for what it does, not what it's called.' Marketing manager:

'Wrong. Customers are like everyone else: impulsive and illogical.' Engineer (thinks): 'Really? Speak for yourself.'

Engineer: 'This modem will make your PC much more effective. It can work at up to 14,400 bits per second. Of course the modem at the other end has to work to V32 bis, and you'll need version 3.2 of the comms software to drive it.' Manager: 'Oh. And what does that actually mean to me as a user?' Manager (thinks): 'I only need to send a few memos. This technology's all getting too complicated, and the IT people can never explain it to me clearly.'

Incidents like these really happen. Think for a moment about similar examples from your own experience. What caused the misunderstanding, and did either party do very much to try to resolve it? As in these examples, did they just think about the real problem and avoid bringing it out into the open? Such incidents give rise to the generalisations mentioned earlier; they become common perceptions. However, one would be wrong to assume that they will apply in every particular situation. Rather, one should view them as pointing to potential difficulties, so that one can be ready to overcome those difficulties as and when they occur. They represent the first aspect of the challenge which this book is designed to address: the diversity of roles and people in the business enterprise, resulting in those four barriers to good communication.

1.2 The challenge for engineers

Look at how this challenge relates to engineers in particular. Consider first what engineering is all about. That will help explain why engineers have their particular perspective, just as other professionals have theirs.

Engineers use technology to solve practical problems. The technology which they use is underpinned by scientific knowledge and principles. Engineers must therefore have:

- factual knowledge about the technology and scientific principles relevant to their branch of engineering
- an ability to understand the context of a practical problem, and then define it in terms which can be understood and agreed by the client
- an ability to use technology creatively to give the most effective solution to the client's problem
- an obligation to do this ethically and in the best interests of the client; this is the basis for trust and a professional relationship

Professionals cannot succeed without a fascination, even love, of the skill which they offer. A musician will love music and be fascinated by the structure and forms which music can take. An engineer will love solving practical problems and be fascinated by the technology which makes this possible. This has led to the jibe that engineers treat things like people, while managers treat people like things. That jibe is not fair to either party but it does highlight a common perception about engineers. They are sometimes seen as: introverted; interested in technology rather than people; cold, logical and calculating; and not very good at human relationships and team working.

There is little psychometric evidence to support this perception of engineers. Engineers spend most of their time being normal people, just like everyone else. As a group, engineers have no startling difference in their psychometric profile (personality characteristics) when compared with other groups. However, there is a difference in the experience which engineers have in their early professional career. This leads to behaviour which other people may label in the way mentioned.

That's important. The problem is not about what engineers are; it's about what they do. It's behaviour rather than being: nurture rather than nature. This fact is encouraging because although we can't do much about our underlying nature, we can do something about our behaviour. And eventually that will change other people's perception of us.

The problem has become more pressing because of the business trends mentioned in the first section. In the old days the engineer's 'client' was perhaps the company's production manager. Engineers tended to work for other engineers. Now, the engineer must deal with a much wider range of people inside and outside the company. He or she is more likely to be involved directly with the company's customer. This requires a wider viewpoint and the ability to work with a customer who often lacks technical knowledge. It is important to succeed because we have seen that business success, and even survival, is determined by the customer.

Thus the first aspect of the challenge arises from the diversity of people inside and outside the business enterprise with whom the engineer needs to communicate. However, the engineer's professional training should provide a good basis for meeting that challenge.

The second aspect of the challenge arises from some of the recent management trends resulting from the customer-led approach to doing business. Here are some of those trends, which impact directly on the practice of engineering:

— Quality is judged ultimately by the external customer. It means understanding the customer's problem and immediate needs, and then meeting those needs effectively. The customer should then be well satisfied; ideally the customer should be surprised and delighted. Total quality management (TQM) is a philosophy which extends that objective to include the internal working of the firm. Everyone in the firm has a customer, whether an internal colleague or an external customer. This concept is novel to many people, who have been used to working just as a cog in the machine. For the engineer, though, it should be second nature. It fits exactly with our definition of what engineers should be able to do, although not necessarily with the scope of every engineer's current job. Thus an engineer's professional training should be an advantage when the firm introduces TQM.

— Business process redesign (BPR; sometimes called business process re-engineering) undertakes a fundamental review of the business processes used within an enterprise. It looks at what is to be achieved from the customer's viewpoint, and then looks at radically different ways in which the required result can be achieved. It usually aligns the business much more directly with the production of the good or service which the customer wants. Again, engineers should be well placed for this. They are used to thinking about production, in this wide sense, and do not want it hampered by administrative procedures and controls which were introduced in the past.

— One specific consequence of BPR is concurrent engineering (CE). This involves the overlapping of stages of design and production which in the traditional project life cycle were strictly sequential. Thus research, design and production staff will need to be involved more closely together. More recently the idea of superconcurrent engineering has brought the customer into this process. Preliminary design work is done with the customer involved, accepting that this will lead to an iterative development of the formal requirement. This approach is all about defining the problem and identifying the solution together. Good communication is central to that.

— That wider approach requires a better knowledge of the principles of marketing and of the commercial environment within which the work is being done. No engineer today will be unaware of the market in which their firm's product has to compete: who the competitors are, what products they offer, and the basis on which their own firm has chosen to compete with them.

— At the same time there has been a loosening of administrative control within companies. Downsizing (or rightsizing) has resulted in flatter management pyramids, with a wider span of control for individuals. The old idea of the manager as a controller of staff has given way to the idea of the manager as an enabler of staff. Thus there is a greater emphasis on teamworking. Groups of individuals are expected to work together to achieve well defined tasks. They must devise their own internal procedures, checks and balances. They must take collective responsibility for the results which they produce. Some engineers are comfortable with this; they may have been used to working within engineering project teams with that kind of freedom and responsibility. Others may be less comfortable, feeling that their job is no longer clearly defined and tidy.

— At the level of the individual, there is a trend to empowerment. This means giving the individual clearer responsibility and the necessary authority to make decisions and use resources to achieve the required results. Most engineers should find this refreshing, and welcome it.

Thus the second aspect of the challenge for engineers arises from these management trends. You may feel that so far you have not been much affected by these trends. If so, be assured that this cannot continue for long! Other engineers are being challenged by many of the trends, and learning to cope with the constant change which results. In any event, the signs are encouraging. Engineers who can work in accordance with the description given at the start of this section should have no great difficulty with these trends in business and management. Indeed they are better placed than many other people in this respect. One thing is clear, though: they will be working more closely with an ever wider range of people. That requires effective communication. For a general management view of these trends in the modern enterprise, see 'Effective business communication: a director's guide' (1994), and Handy (1994).

There is one final aspect of the challenge to engineers which we need to consider. This is the way that it changes throughout a typical career. Figure 1.1 shows a typical career progression for an engineer. Most engineers, when they have

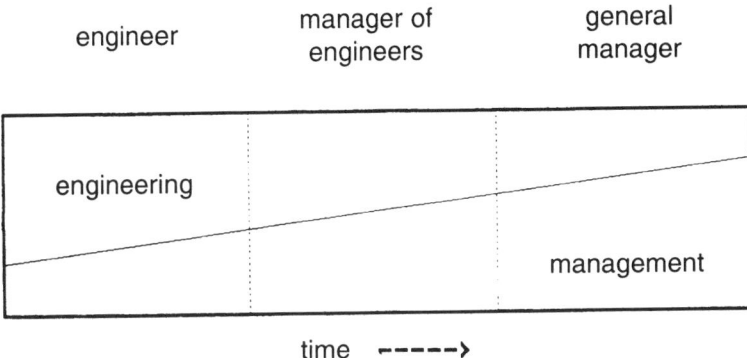

Figure 1.1   Engineer's career progression

completed their technical training, are employed in a job directly related to this. They are probably given specific engineering tasks and are closely supervised by other engineers. Gradually, the scope of their responsibility increases and they take a wider view of the projects with which they are involved. They may get promotion within a design or production department, or take a leading role in the management of engineering projects. They find themselves leading small teams of engineers.

This is the first major career change: from being primarily an engineer doing a technical job, to being a manager of engineers within the context of engineering operations or projects. Many engineers do not make any other major change during their careers; they gain increased responsibility and promotion as they develop within the engineering environment. However, they will certainly need to undertake training and gain qualifications which better equip them for this role.

Other engineers will at some point decide to switch the emphasis of their career away from the technical aspects of engineering and towards general management. What is meant by 'general management'? It is the primary task of those senior people in a business enterprise who, collectively, are responsible for its overall direction and success. They have to take the overview of the business and the environment in which it works. This is often called the 'helicopter view'. The decisions they take are more strategic: they affect the direction of the business over a longer timescale than the operational or tactical decisions of day-to-day business. General managers, when acting in that capacity, must set aside any bias arising from their own background or operational responsibility. They must work together as colleagues, in the best interests of the business as a whole. In this respect they are analogous to the Cabinet in government, taking collective responsibility for the decisions which they make.

Most general managers will have started within a functional specialism and and then been attracted by the wider responsibilities and rewards of general management. In aspiring to that role, engineers are no better or worse placed than other professional groups. However, it does require a conscious personal decision and then a commitment to succeed.

There have been many studies of the skills, or competencies, needed by general managers. Most lists include such terms as intelligence, strategic vision and leadership. Management has been defined as the ability to get things done with and through people. Leadership has an extra dimension: the ability to communicate the vision and inspire people to work towards achieving the vision. Good communication has a central role, both in management and in leadership.

This three-stage model of career progression is of course a simplification. Nevertheless, it highlights the broad stages of development for many individuals. The shift is away from the technical detail of engineering and towards a wider understanding of the business and how people with a wide range of skills make it work effectively. The shift of emphasis from engineering to management is shown diagramatically in Figure 1.1. The shift requires, among other things, a greater skill to communicate.

To draw the threads together, there are three aspects of the challenge for engineers:

1 Diversity of roles and people working within the modern business enterprise
2 Management trends that have arisen from a greater customer orientation of business
3 Wider demands of individual career progression, from a technical role towards a general management role

Each of these aspects requires the engineer to work more closely with a wider range of people and specialisms. Working together is underpinned by communication: the shared understanding of a problem and how to solve it. It also presents the barriers to communication of the last section: engineers being too precise, factual and numerical; nonengineers being too vague; and specialists (including engineers) being obscure. The size of this particular challenge will vary according to the nature of your job, the stage of development of the firm where you work, and your own career aspirations. Two things are certain. First, you cannot ignore it for long. Secondly, you can do something about it. Communication is about behaviour, and with help you can modify behaviour. This book gives you that help.

1.3 Using this book

The aim of this book is to help engineers to communicate effectively with nonengineers in the business context. Figure 1.2 shows its structure. After this

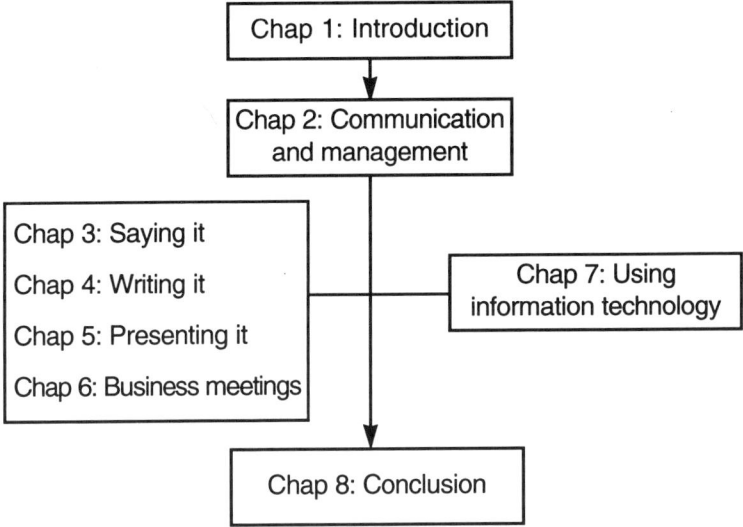

Figure 1.2 Structure of book

introductory chapter you should read Chapter 2. It presents a model of the communication process which underpins the rest of the book. The model shows that communication is not simply a matter of sending and receiving information; it includes feedback, so that the process is essentially two way, and takes place within a context which must be shared by all those participating. Chapter 2 also develops some of the ideas we have mentioned about the business enterprise and the role of management. Overall, Chapter 2 sets the scene for our topic of communication in business. It provides a conceptual basis for the rest of the book.

Chapters 3-6 are an overview of the core skills of communication in business: speaking, writing, presentations and meetings. You may wish to read only some of these chapters at first, depending on how you perceive your own needs.

Chapter 7 is rather different. It deals with the use of modern information technology as a personal aid and across the business. It also looks at the trends in this important area. The emphasis is not on the technology but on how you can use it effectively to support communication between people in the business context. You may wish to omit Chapter 7 at a first reading, or to concentrate on it if IT is beginning to have a major impact on the way you work.

Chapter 8 gives a summary of the points in the main chapters and some advice about continuing to improve your skill as a communicator in business.

Thus the book aims to be comprehensive by dealing with all modes of communication, both direct and indirect (using IT). After Chapter 2 it consists of short chapters on particular topics. You should therefore be able to focus on your particular needs, and it should be more useful for reference afterwards. To make best use of the material you should consider yourself, your situation and your career aspirations. Here are some questions which may help you:

Do you often find there is misunderstanding when you speak with people outside your own immediate area of work? Have you experienced some of the barriers to communication which were illustrated earlier? If so think carefully about the model in Chapter 2 and then read Chapter 3.

Do you have difficulty getting ideas down in writing? Do people come back with questions which show they have not really understood what you wrote? Does your written work read like a transcript of what you might have said in a discussion? Do you find it difficult to marshal your thoughts and structure your writing well? If so, study Chapter 4.

Do you quake at the prospect of a giving a formal business presentation? Are you uncertain about planning and delivering it? Are you uncertain about visual aids? Chapter 5 is for you.

We have all experienced terrible business meetings which seem rambling and inconclusive. Do you feel that you could improve your contribution to meetings? Do you worry about taking a major role such as chair or secretary? Are you unclear what a meeting should achieve and how it should be documented? Chapter 6 will help.

And what about IT? Do you know little about IT but feel you should be using it to better effect? Or perhaps you are aware of the technology but find it difficult to persuade business people to look more closely at the opportunities it presents? Do you have the uneasy feeling that IT is changing the world of business, and you need to be ahead of the game? If so, Chapter 7 should be useful to you.

This represents quite an ambitious agenda for the book. It is also an ambitious agenda for you as the reader. There are two aspects which you need to keep in mind. The first is your objective in using the book to improve your communication skills. This might be to

- improve your performance in your current job
- equip yourself for changes which are taking place in your organisation, or
- equip yourself for career development, to become a manager of engineers or a general manager

A clear objective of that kind should help you to think critically about your present skills and how they need to develop. It should also help you think of practical examples of the issues discussed and ways in which you can apply your developing skill. So, be clear about yourself, your work situation and your aspirations.

The second point to remember is that, after Chapter 2 you can be selective about which chapters to read. However, when you are reading a particular chapter please do so carefully. At several points in the text there are questions for you to consider. Although they are not formal exercises, it is worth considering such questions carefully in the light of your experience. You will find they help you get the most out of the ideas which follow.

This is not primarily an academic textbook; it is a practical handbook. However, you might wish to read more about particular topics. The 'References, and further reading' list at the back will help. Each item has a short abstract, and most are referred to from the main text. This will help you select further reading if you wish.

You want to be a more effective business commmunicator. By reading this far you have shown that you mean business.

Chapter 2
Communication and management

'Reading maketh a full man; conference a ready man; and writing an exact man. And therefore, if a man write little, he had need have a great memory; if he confer little, he had need have a present wit; and if he read little, he had need have much cunning, to seem to know that he doth not'
from 'Of Studies' by Francis Bacon (1561-1626)

2.1 Introduction

This chapter provides the foundation for the rest of the book. It introduces some concepts that are used in later chapters to discuss the individual methods of communication in business. It is therefore important to get a good grasp of the ideas before moving on.

Section 2.2 deals with human communication. This is a fascinating topic, of which we all have a great deal of experience. Much of our time is spent in communication, whether formal or informal. What we need to do is understand the general process which operates, and its practical limitations. We need some general principles which will guide us in our attempts to communicate more effectively.

Section 2.3 attends to the business enterprise, picking up some of the ideas from Chapter 1 and identifying the increasing importance of communication in making business work effectively. Section 2.4 looks in more detail at the crucial role of management and how it is underpinned by the use of information. This gives a framework for understanding the purpose, aim and scope of any particular business communication process. That's important: business is purposeful, and

12

business communication must support that purpose. It is not just idle or irrelevant chatter.

As with the other main chapters in the book there is a concluding section which summarises the main ideas. This should help you check your understanding before you move on.

2.2 What is communication?

It is difficult to define exactly what is meant by the communication of information between people. To do so with scientific precision is impossible, and the attempt is frustrating and not very illuminating. Instead, consider some simple but practical definitions:

- Communication is the transfer or sharing of information
- Information is human significance associated with particular facts or situations

There are several features implicit in these definitions which accord with our everyday experience of human communication:

- There is a sender, and at least one receiver, of the information
- More often than not, the interaction is two way. Two people alternate the role of sender and receiver. Moreover, even when communication is mainly from the sender to the receiver there is nearly always some mechanism of feedback the other way. This may simply be to question, express surprise, or even to ignore the attempted communication
- This more balanced relationship between the communicating parties is reminiscent of the other meaning of the word: sharing or communion
- The importance, relevance and value of information is subjective: it depends on the interests and circumstances of the recipient. If you are told you have a pay rise, that is very interesting. If you are told that a colleague has a pay rise that is interesting, but in a different way. If you are told that someone in a different company has a pay rise, that may be of very little interest to you.

We are involved in communication during much of our waking hours. How many processes of human communication have you been involved with today? Think carefully. There are the obvious ones, like speaking to people at home, or on the journey to work, or at work itself. But what about listening to the radio, watching TV, reading the newspaper or reading this book? What about that packet of breakfast cereal you stared at disconsolately: what sort of communication was intended by those pictures and text? And why did you choose to dress the way you did? Does that send a message to others?

The way we interact and communicate in everyday life is complex. The academic study of such communication has had two main threads:

- The static view, which emphasises the meaning (or significance) attached to objects and signs. The science of signs, called semiology, considers the relationship between the sign itself, the object or idea which it represents, and the interpretation or meaning in the mind of each of the various users of the sign. The sign itself has an objective reality, and so may the object it represents. But the meaning for different users may vary. This is the root of ambiguity and misunderstanding. A business logo is an example of a sign. You have probably encountered several of these already today. If you look around, you can probably see several more now. For example, the publishers have placed their logo on this book. One business logo will rely on an image, another on words written in a distinctive way. In any event, signs may have different meanings for different people who see them. This is important in cross-cultural communication.

— In contrast to this, the dynamic view of communication emphasises the process by which information is transmitted from a sender to a receiver. Human speech is a good example; it is transient and ephemeral. Written communication is often hasty, but at least in principle we can pause, reflect and re-read in a way which is not possible with speech. In poetry, for example, the words can often be interpreted in several different ways. The deeper meaning is sometimes not apparent at the first encounter. The pace of modern business life means that communications must be as simple, direct and clear as possible. It is no good relying on a recipient 'reading between the lines' to get a subtle meaning; make your real message clear! Speed, then, is not the only virtue in business communication. Indeed, it is interesting to consider what the overall impact on management would be if we were only allowed to communicate slowly, in rhyming couplets of poetry! Poetry has been defined as 'the right words in the right order'; that principle should apply to any well-crafted writing.

If you wish to explore the theoretical approach further I suggest these references: Cherry (1978), Corner and Hawthorn (1980), Mellor (1990), Shepherd (1971) and Whittick (1971). These two models of the human communication process highlight two important features: it may be ambiguous (so that different people get a different message) and it may have multiple levels of meaning (so that different people will identify with one or more of those levels). Both these facts can lead to misunderstanding. In business communication, one must take care to reduce the danger of that.

It's worth thinking about the various ways in which we can communicate, and which are the clearest. Figure 2.1 is an attempt to draw the semiotic and process models together. To communicate, the sender has to make some movement. The left-hand column of the Table distinguishes large-scale movements (like waving an arm) and small-scale movements (like moving the lips). The receiver may perceive the sender's movement directly; this is shown in the upper part of the right-hand column. However, the receiver may perceive a consequential (but different) stimulus arising from the sender's movement. This is called indirect communication, as shown in the lower part of the right-hand column. Writing is an example of indirect communication; the sender moves a pen, or types on a keyboard, but the receiver perceives not the movement itself but the marks on the paper or screen which result from that movement.

Sender's output movement	slow (information)	rapid (communication)	see	hear	feel	
large-scale — dress			x			DIRECT
heraldry			x			
posture		dance	x			
		semaphore	x			
gesture						
small-scale		non-verbal communication	x			
face	expression					
vocal tract						
time	shout	speech		x		
pitch	whistle	music		x		
hands						
direct — handshake		deaf-blind			x	
handsign		deaf	x			
indirect						INDIRECT
marker — image						
colour			x			
tone			x			
line			x			
drawing		pictogram	x			
calligraphy		write/read	x			
symbols		notation	x			
keyboard						
one key		morse code		x		
multikey		typewriters	x		x	
musical instrument		music		x		
keyboard						
stringed						
wind						
percussion			see	hear	feel	

Figure 2.1   Human communication

The receiver perceives the sender's movement, or the effect of it, with one of the five senses. Seeing, hearing and feeling are the three cases of practical importance, as shown in the right-hand column.

The final classification shown in Figure 2.1 is speed. This is reflected in the subdivision of the central column. The transfer of information may be slow because of physical limitations or because the method is essentially static. This is where the semiotic model of communication is important. If the transfer of information is continuous and fairly rapid, then the process model of communication is important.

Consider some of the examples shown in Figure 2.1:

- Dress, heraldry and posture are essentially slow, large-scale methods of communication. The modern form of heraldry is, of course, the corporate logo used on a tie, badge or other article of clothing. The totally static version of these methods of communication is sculpture.
- Speech is the most important method of direct communication. It is usually complemented by gesture and posture, which form nonverbal communication (NVC). Remember, though, that nonverbal communication is not available to a blind person, or someone listening over a telephone. One of the advantages of a videoconferencing call is that it transmits most of the nonverbal communication which is lost during a conventional telephone call.
- Writing is the most important method of indirect communication. When done with a pen or other marking implement it may be slow (calligraphy, the art of beautiful writing) or rapid (as in normal handwriting). When done with a keyboard, the sender and receiver can take full advantage of modern electronic communication facilities.

Figure 2.1 shows that the final stage in the communication process is the receiver's perception of a stimulus. This is by seeing, hearing or feeling. Perception is a complex process, involving not only the sense organs themselves but also the processing within the cerebral cortex of the brain. Errors can occur at either stage. For example:

Seeing: Optical illusions are well known. Figure 2.2 shows six black objects, yet the brain interprets the stimulus as an opaque white triangle overlaying three black

Figure 2.2   Optical illusion (1)

disks and a black-line triangle. The white triangle almost looks 'whiter' than the surrounding paper, but this is wholly an illusion. In Figure 2.3 some familiar text is often misread because the eye is deceived by the layout within a triangle. Such ambiguities of perception are not just curiosities; they are important when considering the presentation of graphical information in business.

Figure 2.3   Optical illusion (2)

Hearing: We may simply fail to hear sounds, because of a noisy environment or because we are preoccupied with some other task. Even when we do hear, the interpretation or perception of sounds depends on the context in which they appear. It has been found that in short timescales, and depending on the context, sound may be perceived as silence, or silence may be perceived as sound.

Feeling: There is an ancient paradox of putting one hand into a bowl of hot water, the other into a bowl of cold water, then putting both into a bowl of tepid water. The hands then perceive the same water in a different way: one as hot and the other as cold. The acuity of the sense of touch varies widely over the human body, the tip of the tongue being the most sensitive. It is possible to convert visual stimuli into tactile stimuli, to enable blind people to 'see' a simple image. Modern IT is making important bridges of this kind, so that people with disabilities can become involved in communication processes previously not possible for them.

You may be interested in reading further about some of the specific methods of human communication. See Barlow (1990), Gregory (1990), Jackson (1981), Morris (1994) and Wood (1975). In any event it is worth reflecting on the everyday communication situations you find yourself in, and considering how the semiotic and process models apply. What are the lessons for business communication?

Now focus on the methods of communication used in business. For practical purposes, one can identify four main modes of handling information: text, data, voice and image. They are used in various combinations, and modern IT is drawing them closer together. The two commonest examples are text supported with image (like a management report, or this book), and voice supported by image (as in a television link). It depends on the business situation whether the supporting mode is important or not. Thus diagrams can avoid the need for a lot of written material. Sometimes seeing the person you are talking to is very important; NVC is an important aspect of interview or other negotiation situations. You must judge whether it is worth the time and expense to arrange a face-to-face meeting or videoconferencing link in such situations, rather than accept the limitations of a telephone call.

Think for a moment how much you use the four modes of handling information in your work. Which of them is most effective, in which situation? Buzan (1988) suggests that we spend between 50 and 80% of our waking hours communicating. That communication time breaks down in these approximate proportions: 45 % listening, 30 % speaking, 16 % reading and 9 % writing. The amount of formal instruction we receive is usually in inverse proportion to those figures; we have paid most attention to the skill which we use least.

That gives some idea of how we spend time on communication, but what about its effectiveness? Written communication, on which we spend least time but for which we receive most training, can have the greatest lasting impact or reach the greatest number of people. Considering ordinary spoken conversation, there are three distinct channels in operation:

- verbal: the words spoken
- visual: distance, environment and gesture (nonverbal communication)
- paralinguistic: variations in speech, for example pitch and loudness

Psychologists estimate that only about one-third of the social meaning of a conversation is carried by the verbal channel (the words). Considering how emotion is conveyed, the words are even less important: about 7% of emotion is carried by the verbal channel, 55% by the visual channel, and 38% by the paralinguistic channel. These figures probably accord with your experience of social conversation, but are they really relevant to business communication? Yes they are, because

— even when we are trying to convey a factual, objective message the words convey less than half of the message
— to convey our sincerity, commitment or other emotional aspect, the visual and paralinguistic channels are far more important than the words. Thus a business presentation, face-to-face with senior managers can have a much greater impact than the written report which it supports. If the presentation establishes your credibility, those managers will read your report with closer and more informed attention.

Some of these ideas are drawn together in the form of a model of the business communication process shown in Figure 2.4. Most engineers will recognise the central part of the diagram. It comprises the classic process model of communication pioneered by Shannon and Weaver (1947). The sender encodes the message into a form suitable for transmission along a communication channel. The channel may be disturbed by unwanted interference called noise. The receiver decodes the message from the channel and hopefully recovers the original message.

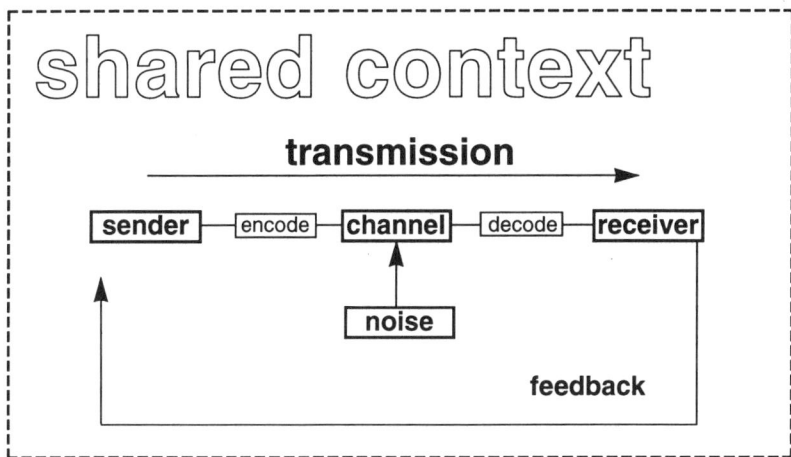

Figure 2.4   Model of communication process

So far the model is one-way: from sender to receiver. However, there are two other important features of the model:

● Feedback is the immediate or delayed response which the receiver, or receivers, of the message may choose to make. This feedback may use a similar or different type of communication channel back to the sender.
● Shared context is the common understanding, shared by sender and receiver, about what is being communicated and why. For example, a simple requirement when communicating in words is to know the language being used, to the necessary level of proficiency. More broadly, the partners in the communication process must share a general understanding of the topic under discussion. This will influence, for example, the use of technical or other specialist terms.

This extended model of the communication process enables us to place individual acts of communication within the business context in which they take place.

There is another aspect which can be borrowed from the classic work of Shannon and Weaver (1947). This is the measure of effectiveness of the communication process, shown in Figure 2.5. The diagram shows three levels. The first level (technical) is the most detailed, and relates to the central part of the Figure 2.4 model. The second (semantic) is concerned with the effective transfer of meaning from sender to receiver. Feedback, leading to a dialogue, may be the only way to achieve this. The third level (effectiveness) is concerned with the wider shared context of the two parties.

To illustrate that for the case of a written management report:

● At the technical level, the report must be smart physically. The pages should be adequately bound together, with the title visible. The text and any graphics must be clearly legible. The report must be submitted on time.
● At the semantic level, the report must convey the intended message to its likely

Level A -   technical level

How accurately can the symbols of communication be transmitted?

Level B -   semantic level

How precisely do the transmitted symbols convey the desired meaning?

Level C -   effectiveness level

How effectively does the received meaning affect conduct in the desired way?

Figure 2.5   Effectiveness of communication process
Source: Shannon, C. E. and Weaver, W. (1949)

readers, taking account of their own knowledge and familiarity with the problem being discussed.
● At the effectiveness level, the report should evoke the intended response from the readers. They should feel that they have been provided with what they asked for, or needed; that it helps them understand a particular management problem or situation; and that they are better equipped to solve it.

Now consider how the communication model and the levels of effectiveness apply in a broader way. Take speech first. It is the most familiar method of communication in our personal, social and business lives. It is often informal and spontaneous. This is quite the opposite of the management report just considered! It is worth distinguishing two situations for spoken communication, depending on how many recipients there are:

One-to-many: If you are addressing many people at once you are making a presentation. It is not possible for each member of the audience to give you their feedback at once, although some may try to by shouting, heckling, or posing uninvited questions. Some people may go to sleep; that provides you with a different kind of feedback, and a challenge too. However, most of the feedback is likely to come at the end of the presentation when people may ask questions or come to speak to you individually. Ultimately, the feedback may be whether you are ever invited back to address that audience again. During your presentation you will have to make some assumptions about the shared context between you and the audience. You can safely assume that they were made aware of the topic of your talk beforehand, and would not be there unless they had at least some interest in it. It is more difficult to judge the level of their knowledge so that you can speak to best effect to the majority of those present. It may be fun to enter a dialogue

with the experts in the audience, but that is not much use for the majority
who may have no idea what you are talking about.

One-to-one/few: If you are speaking to one other person, or a small number
of people together, then you are engaged in conversation. This is quite
different from a presentation. For one thing, the feedback is likely to be
immediate. If people do not understand what you are saying, or disagree,
then they will tell you very quickly. We have developed elaborate conventions
of using nonverbal communication (NVC) to change our turn to speak in
a conversation. We signal by facial expression and verbal intonation when
we are about to end our contribution, so that the other person can be ready
to respond with theirs.

Both those speaking situations use a verbal and a visual channel between sender
and receiver. This is voice and image, in our classification of modes of information
handling. With modern technology both the situations (presentation and
conversation) can take place at a distance, using videophones or videoconferencing;
these are discussed in Chapter 7. Written communication includes written words,
and still images. There are two equivalent situations to those just discussed:

One-to-many: Books, magazines and newspapers are examples of this type
of written communication. We may buy a book because the title attacts us,
because we know the author, or know their previous work and reputation.
We may buy magazines and newspapers because we are interested in the
topics they address, and have come to enjoy the style and viewpoint adopted.
Writers of such one-to-many communications have to make assumptions
about the readers: the interests, shared context and likely level of knowledge.
This is similar to making a presentation to a large audience. For the readers,
though, the situation is different. They have no opportunity for immediate
feedback, including questions of clarification. They can write letters to the
editor or author, but this is cumbrous and slow. Collectively, they can
withdraw their support from the publication, so that it ceases to be
marketable. Thus feedback is still available, but it is slow and difficult.

One-to-one/few: The personal letter is a good example of this type of written
communication, although we do not write as many as we used to. In business,
the office memo (whether on paper or by e-mail) and the management report
are common examples. The difference is that we now know the reader or
readers as individuals and can take account of that in the way be prepare
the written communication. Moreover the readers usually have opportunity
for quite rapid feedback to you as the author. They may react with surprising
speed and vigour.

The next three chapters deal with spoken and written communication in more
detail. The distinction between the one-to-many and the one-to-one/few situations
will be an important one to consider.

So much for the mechanisms or channels of human communication. Now
consider the limitations which those channels impose. It is obvious that one cannot

take in, process or transmit information at an indefinitely high speed. What are the practical limitations on our performance?

As an engineer you will like to quantify things where you can. You probably support the old dictum that 'if you can't measure it, you can't manage it'. It is not wholly true, but one thing is certain: when we can measure, we should measure.

Here are some representative figures for common communication activities:

Writing: A typical longhand writing speed is 25 words per minute (wpm). You can increase this, proably at the expense of neatness and legibility. If you learn the rudiments of shorthand you will soon be able to take notes at about 100 wpm. However, it is unlikely that anyone else will be able to read those notes. Sometimes this can be an advantage, for example taking confidential notes in a meeting which you do not wish your neighbour to read. However, shorthand is a less useful skill than it used to be; keyboards and modern IT are more effective. Your typing speed is limited by the method you use. If you are a two-finger typist then you will not achieve a lot more than your longhand writing speed. Today many people find it worthwhile to learn touch typing: using all eight fingers and two thumbs, and rarely looking at the keyboard. At first you will be slower, but should soon reach a speed of 40 wpm.

Reading: People commonly read at about 240 wpm. By controlling eye-movement and other techniques it is possible to achieve twice that speed (Buzan, 1989). In business, many people find they have to use speed reading for some documents, simply to keep abreast of the all the paperwork that comes to them. There is much material which you do not need to read in detail. Unfortunately, you often do not know that this is the case until you have given it a quick preliminary read! In any event when approaching a book or other large document it is a good idea to skim through it quickly so that you know the main theme and the structure of it. You can then focus on any detail which you need.

Speaking: A typical speed in conversation is 150 wpm. In a formal presentation you should speak more slowly, at about 120 wpm. In lively conversation you may exceed 200 wpm.

Listening: Most people can listen faster than they can speak, perhaps up to 400 wpm. Whether they absorb the information is a different question, discussed below!

Remember that these figures relate only to the channel of communication, not to the encoding and decoding of the useful information. Think again about the model of the communication process, shown in Figure 2.4. For example, if you are drafting text you normally take longer to think about the content and compose the text in your mind than you take to write it down. My experience in writing books and educational material is that it takes two hours to draft one thousand words of text. This is even after assembling the material and planning the structure of what you want to write. This corresponds to an average of just over 8 wpm.

You could therefore argue that longhand writing at 25 wpm is not going to delay the process. This would be the case if you composed and wrote steadily, perhaps like a Victorian novelist. However, the process of creative writing (thinking, then drafting chunks of text) is not smooth or continuous. Spells of thought are interspersed with periods when the words come quickly. It is then that you need to record them fast. In practice, I find that 30–40 wpm at a keyboard prevents any holdup in the process. Longhand writing at 25 wpm would be a serious hindrance, and in any case would mean that someone would have to type the draft at a keyboard later. So, let's take a figure of 8 wpm for generating business-level material.

Thus for complex matters, such as we often deal with in our work, the limitation is not the channel of communication but rather our mental capacity. It is therefore helpful to see how information theory can help us to interpret the speed of a text channel. Ignoring punctuation, text comprises a selection of 27 signs: the 26 letters of the alphabet plus a space. If the 27 signs were all equally probable, then the information content would be 4.75 bit/sign (this is the logarithm of 27, to base 2). You recall that the measure of information, at this technical level, is the bit, which represents a choice between two equally probable alternatives.

Taking account of the actual probabilities of the 27 signs in written English, the information content reduces to 4.03 bit/sign. However, we know that letters cannot be chosen independently of previous letters. For example u always follows q, and x is unlikely to follow z. Taking account of the preceding symbol, the figure drops to 3.3 bit/sign. Taking account of all preceding letters and words, and the context of meaning, the figure drops to roughly one bit/sign, or 5 bit/word. Speaking at 150 wpm, this corresponds to about 12.5 bit/s. Remember this figure only relates to the words. We may also be presented with information in other ways, such as nonverbal communication (NVC). It is much more difficult to quantify those, even though we have seen that they can be even more important than the verbal channel.

What are the implications of all this? Well, it turns out from studies of perception that people can only absorb and act upon information at an effective rate of about 5 bit/s. This figure is remarkably independent of which channel (or combination of channels) is used for the stimulus, and what kind of response is required from the recipient. For example, a person playing fairly simple music, at sight, is processing information at about 5 bit/s.

Figure 2.6 summarises all these figures diagrammatically. The precise figures do not matter very much, but their relative sizes do. Note that:

- We can only generate new structured information at an average speed roughly equivalent to 8 wpm of English text.
- We can only absorb new structured information at a speed which is roughly equivalent to 60 wpm of English text.
- We normally speak at about 150 wpm. Most of what we say will therefore not be absorbed by the recipient.
- We can listen at much higher speeds. This further reduces the fraction of the information which we can absorb.

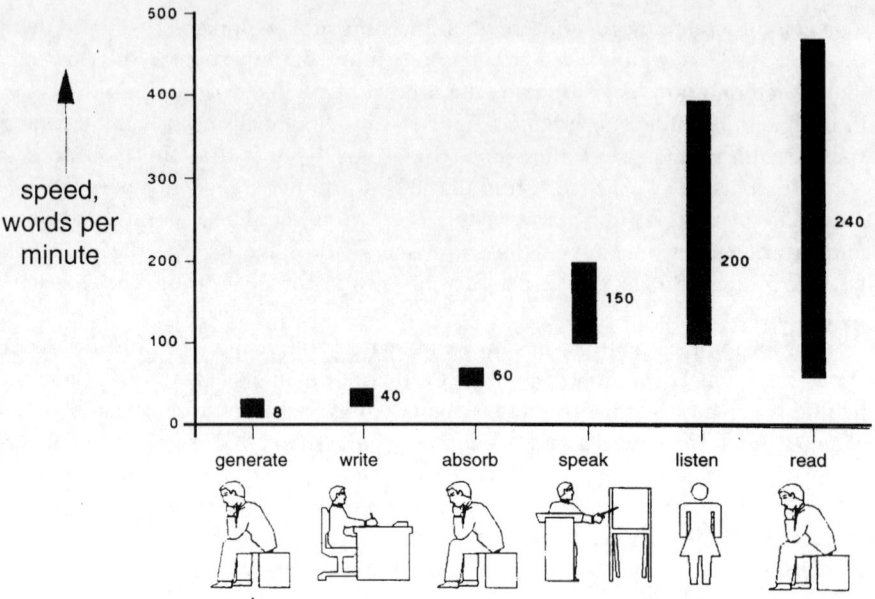

Figure 2.6   Communication speeds

- We can write at about 25 wpm (longhand) or 40 wpm (on a keyboard).
- We normally read at about 240 wpm. This is very much faster than we can absorb new information effectively.

These figures are only meant as a rough guide. However, it is worth doing some measurements and experiments to find out typical figures for your own performance. This will help you to

— Plan major tasks like report writing and business presentations. It is useful to know how many words you write on an A4 sheet of paper in your usual drafting format, whether this is handwritten or on a word-processor. Develop the skill of judging the number of words on a page of text in the formats you regularly encounter.
— Decide whether, for your style of working, it is worth learning to touch-type or acquire some other skill which will improve your overall productivity.
— Decide on the right balance between speaking and writing in your work. The technique of dictation (to a secretary or via a minicassette recorder) can save you time for straightforward documents. You could even use the telephone, or a modern voice-mail system instead! For complex documents, though, you may find your speed of composition is much less than the speed of dictation. That can waste a lot of time.

The broader implications for business communication are

— We must structure our message (whether spoken or written) so that the key points are repeated and emphasised. Of course, this can become stilted or

annoying: there is a special word (perseveration) for protracted and spontaneous repetition.
— When making a presentation (at perhaps 120 wpm), this repetition or redundancy must amount to half the total message. Otherwise, listeners are bound to lose some of the content.
— When reading, we must consciously adjust our speed. To fully absorb closely argued text one must slow down to an average of about one quarter of normal speed. This will usually be done by a combination of reading slowly and rereading key passages until it is understood fully. To absorb the general message and flavour of the text, one can read at normal speed, or even faster. However, for those who cannot spare the time to do even this, one must provide a separate overview summary which they can absorb quickly and effectively. This flexibility to suit different readers is an important advantage of written communication in business.

2.3 Business enterprise

Chapter 1 said that the aim of a business enterprise is to satisfy its customers, and then its other stakeholders. It identified some of the trends in management that are occurring in the 1990s. Key developments are: total quality, business process re-engineering, concurrent engineering, the market-led organisation, right-sizing, teamwork and empowerment. The key underlying issue is the need to satisfy the actual and latent requirements of customers, and to do this quickly, effectively and profitably. Organisations are becoming flatter, looser and more responsive to help achieve this. The product offered to the customer might be a simple service, like cutting hair. If the product involves a tangible good like cement or washing machines, the product is not just that tangible good. The product must comprise the good plus a service which delivers it effectively to the customer. In today's business environment the product can no longer just be the tangible good. For example cement manufacturers mustn't just offer cement; they must offer a cement service. And they must do this in the face of competition. This trend is one reason for the shift in the basis of developed economies away from manufacturing and towards the service sector.

This business situation is shown simply in Figure 2.7. The customer is the most important person. That is the first principle of business in general, and of marketing in particular. Customers are not particularly interested in the business as an organisation; they are interested in the product offered. And the product comprises a service and sometimes a tangible good to go with it. The customer will look at the characteristics of the product and compare them with the characteristics of a competitor's product. One needs to ensure that this comparison results in the customer buying our product rather than the competitor's.

The three overall characteristics which define a product, in this wider sense, are

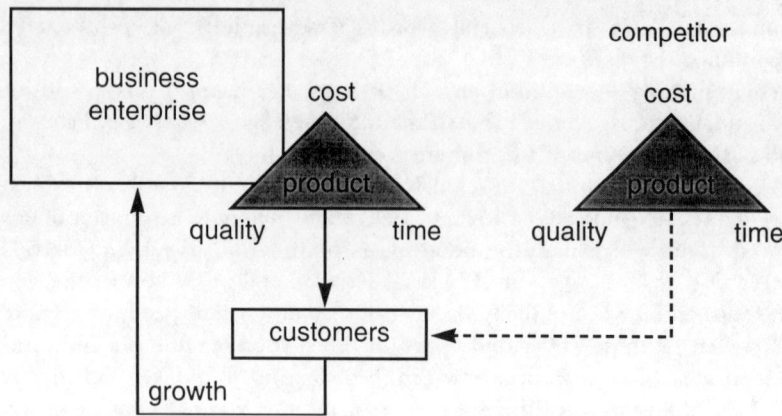

Figure 2.7   Business, products and customers

Cost: the price you ask for your service. This is the characteristic which is most directly affected by competitive pressure. It is also the one which most immediately affects the cash flow and profitability of your business.

Quality: how fit your product is to meet the requirements of the customer. For many years, we have produced tangible goods to meet specifications. One must now meet tighter specifications, and do so more consistently (a 'right first time' policy). Increasingly, one must also meet specifications for service quality. This may be because our business is accredited against international standards like BS 5750 or the ISO 9000 series. It may also be because the service aspect of our particular product is defined by a service-level agreement (SLA). This is the case, for example, for facilities management (FM) companies which offer IT services.

Time: this includes the speed with which you improve your product, or bring new products to market, as technological and other developments make that possible. It also includes the speed with which you can respond to a customer's order, once it has been placed. This may involve customising the product, to meet the precise needs of that particular customer on that particular occasion. It includes modifying your product to meet the evolving needs of your customers over a period of time; this may include interacting with them about the possibilities for the future. The trend is towards long-term partnerships between suppliers and customers, to make such forward planning possible.

To deliver the total product (good plus service) in this way, the business enterprise will need an information system, a manufacturing system and a logistics system. Those systems, and the people who comprise the most important part of them, must work effectively together. To do this they must communicate! So, the implications are

— Wherever you work in the business, you must be aware of the total business situation. In particular you need to understand the business situation as depicted in Figure 2.7; you must understand customers and their needs, the products which you and competitors are offering, and the basis on which you are trying to make your product the best.

— You may need to be involved in discussions with customers. The firm needs a closed loop with its customers if it is to understand them and respond to their immediate and evolving needs. Engineers who were previously in R&D or manufacturing jobs buried inside the firm, are now finding themselves alongside marketing staff and general managers in discussion with customers. They may even find themselves placed for a short while in the sales area, to give them direct (and offer bitter) experience of dealing with customers. In such situations, you need to think in business terms, and communicate effectively outside your own professional specialism.

— Even inside the business, you will be dealing with a wider range of colleagues. This is necessary if the shared understanding of the business situation is to be met by a co-ordinated information, manufacturing and logistics system.

— You may become involved with other stakeholders in the business: outsiders who have an interest in the business. Sometimes that interest may be benign, and sometimes it may be hostile. Janner (1988) gives useful advice on dealing with the public in presentations and meetings.

There is one more trend in business which we should note, because it has a direct affect on communication. This the trend towards international business, which is obvious to even the casual observer. There are some special barriers or difficulties when we are involved in international business:

> *Time:* There may be significant time-zone differences, and even a date-line difference, between the various locations involved. There are delays in physical transportation, compared with purely national operations. There are delays in the post and other communication systems. Even electronic communications may be slower, or of lower capacity or quality.
>
> *Distance:* The problems of geography and distance result in higher transport costs. These may be made worse by a limited transport infrastructure in the countries concerned. Often the only options are sea and air transport; these may seriously constrain the products which we can sensibly offer in overseas markets.
>
> *Complexity:* The factors already mentioned make international business more complicated than national business. But there are even more. We may have to deal with a variety of nations, political systems, governments, laws, ethnic and cultural groups, languages, natural resources, infrastructures, markets, suppliers, customers and competitors.

These developments are important. Engineers should take a particular interest in them because technology can help to overcome the three sets of barriers. This involves appropriate information, manufacturing and logistics systems to support the international business policy. For further discussion of this, and current examples, see Silk (1995).

There are very few businesses left which are wholly national in their operation. At the very least, firms buy raw materials from abroad or export some of their products. There are probably no businesses involving engineers which are not now international in a broader sense than that. Thus we must all come to grips with the customer-oriented business, and do so in an international arena. Some of the communication implications of the international aspect are as follows:

— The need to deal with diverse languages and cultures. One may need to speak more than one language, or have rapid access to people who do. One must be sensitive to the greater risk of ambiguity and misunderstanding in cross-cultural communication. This is especially so for social custom and nonverbal communication when one meets people from abroad or travels overseas.

— The need to overcome, or perhaps exploit, time-zone differences. There is often a window of only a few hours when one can interact directly with people abroad. Modern communication messaging systems such as fax, electronic mail (e-mail), computer-mediated communication (CMC) and voice messaging can be a great help in this situation. One can sometimes use those to achieve a 24-hour style of working, with the work being passed between colleagues around the world so that each is working only during normal daytime hours. Manufacturing firms use linked computer-aided design systems to achieve a faster design capability so that they can respond more quickly to customer requirements. Financial institutions similarly work around the globe and around the clock.

— Using technology based systems to overcome communication difficulties. There is a widening range of telecommunication options available within and between developed countries. This often highlights the poor telecommunications infrastructure in less developed countries. However, it is possible to use satellite based systems to link business locations where the local infrastructure is inadequate.

2.4 What is management?

Section 2.3 considered what the modern business tries to do. The key task is to respond effectively and profitably to the evolving needs of customers. The international aspect adds both barriers and opportunities for this. It showed the implications of all this in terms of the increased need for communications inside and outside the business. This section comes closer to what individual business managers have to do, and how that work is supported by information. This will help define the role of individual acts of communication, and in particular, to define their purpose, aim and scope.

What do managers do? Think about that question before moving on. There is no agreed definition of the term management, so that confirms that the question is not simple. However, let's again use a simple working definition: managers direct resources to achieve business results. In that sense we are all managers, even if we only manage our own work and the resources we need to do it. The key ideas of the definition are:

- Managers don't do it all themselves. They plan and co-ordinate the work of themselves and others.
- There are many kinds of resource available: people, money, material (including all tangible assets), energy, time and information. They all need to be managed. This is discussed in Silk (1991).
- The work is purposeful. It is directed towards achieving stated business objectives. If the business objectives are not stated and understood work will be ineffective.

The management guru Peter Drucker has said that the key functions of a manager are to take decisions and initiate action. Managers can only do this if they have appropriate information on which to base their decisions. The whole process can be viewed as a continuous cycle, where managers learn and organisations develop. This management learning cycle is shown in Figure 2.8. The diagram shows a

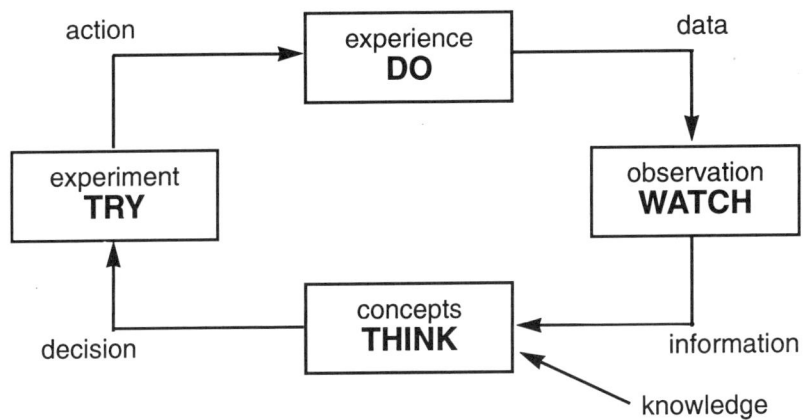

Figure 2.8   Management learning cycle

four-stage cycle. This is summarised by four words: do, watch, think and try. The other words in the boxes are a bit more specific about those activities. The cycle is underpinned by a sequence called the information value chain: data, information, decision and action. The whole cycle works like this:

> *Do:* In the course of our work we experience real events. Some of these events relate to the business, some to other people, and some to ourselves. We are exposed to a myriad of detailed facts about the real-world environment. These detailed facts are called 'data'.
>
> *Watch:* To learn from experience we must consciously observe what is happening. It's no good having your mind on something else (even the next thing) all the time. We have to look critically and selectively at the mass of data to which we are exposed. We focus on the data relevant to the

situation we are trying to manage, or to the problem we are trying to solve. We analyse the data to try to find its significance. In the case of quantitative (measured) data, we may do some statistical analysis. In both cases the result is 'information', as defined at the start of Section 2.2: human significance associated with particular facts or situations. Watching therefore involves critical observation, analysis and interpretation of the facts. It turns data into information.

Think: We must then think about that significant information. We may need to bring in 'knowledge' from outside the immediate situation to help us understand it. Knowledge may be in the form of the previous experience or professional knowledge of ourselves or relevant experts. It can be knowledge recorded in the files or archives of the company, or in a specialist book or journal. It can be knowledge gleaned from our customers or other people or organisations outside the company. In each case knowledge is relevant, distilled information brought in from outside the immediate situation. By relating this knowledge to the information from the immediate situation, we can formulate concepts and ideas about how things could be improved. This might be working out the options for solving a problem, and then choosing the best one. In that case the output of the thinking process is a 'decision'.

Try: Once a decision is made we must decide how to put it into effect. There may be many people and other resources involved. For example if the decision is to develop and market a particular product within one year, a complex plan will be needed. When the plan has been made, usually with the participation of all those involved, then it can be put into 'action'. This usually requires a large number of individual instructions, harmonised as part of the larger plan. As this action takes effect, it results in new experience. The cycle then starts again: we must watch and think; we may also need to try adjustments to the plan. The total process of the learning cycle then accords with another definition of the task of managers: plan, execute, monitor and control.

Look in closer detail at the information value chain, because that is where business communication has its most important role. If the management learning cycle is unrolled into a linear form, one gets the diagram shown in Figure 2.9. This shows the sequence over the course of time. We start with an old situation, and the data about it. We go through a process of change: watch, think and try. The resulting actions give rise to a new situation for which we must repeat the process.

This way of looking at it has two advantages:

- It fits well with a popular metaphor for the process of managing change: unfreezing a situation, changing it, then refreezing it. This is reflected in the stages shown at the bottom of the diagram.
- It focuses on the information value chain which supports the change process shown in the middle. It starts with a lot of data about the old situation and ends with detailed action points which will initiate the new situation.

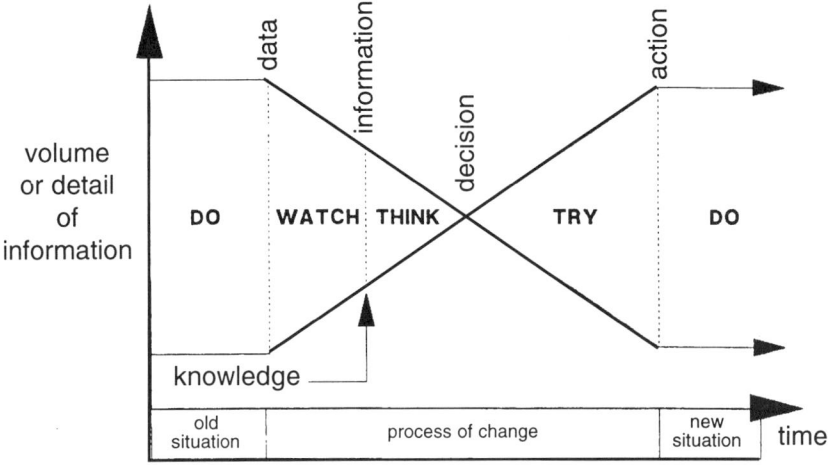

Figure 2.9   Information value chain

Notice how the volume of information first reduces and then increases. It converges from data to decisions. It then diverges from decisions to action points. The decisions are the pivot or turning point. Think about some of the decisions which have been made in your work situation:

● How were the decisions arrived at? How does the 'converge' part of the model fit with what actually happened?
● How were the decisions recorded?
● How were the decisions put into effect? How does the 'diverge' part of the model fit with what actually happened?

I hope you decided that decisions were properly recorded. For important ones, this should usually be in writing. For minor ones it may be verbally, with the full involvement of those affected. In either case there is a danger of mis-understanding, or omitting someone who ought to know or be consulted. Has this happened in your experience?

Figure 2.9 can be used as a model for the process of communication in business. For any substantial business communication, like a report or presentation, you need to decide (and state)

● *Purpose:* the reason for making the communication. This usually relates to solving a management problem.
● *Aim:* what the communication itself tries to do. This is usually to contribute in a specific way to solving the problem.
● *Scope:* how much of the information value chain the communication deals with.

The purpose defines the context; the aim defines the contribution within that context; the scope defines how far along the information value chain (or round the management learning cycle) one intends to go. The aim should make a clear contribution within the overall purpose. The scope should be consistent with the aim. Purpose, aim and scope are important. They help us to structure and prepare effective management communications. This then completes a useful acronym to remember: PASS: purpose, aim, scope and structure.

To illustrate this, consider some written management reports. Here are some business situations which could give rise to a report being written:

1   Move head office out of the capital city. As a first step information is needed about possible locations and costs.
2   A quality problem with a supplier. We don't really know how to deal with this.
3   A morale problem that has arisen since reorganisation. Some specific ideas are needed for improving things.

In each case the management report will not solve the whole problem; life is not that simple! Rather, the report must make a well defined and clear contribution to solving the problem. The purpose of the report in the three cases might be to

1   provide information which will help decide where to relocate the head office
2   define the supplier quality problem more clearly and suggest how one might solve it
3   define the morale problem and suggest how one might solve it

Notice that the problem situation is a statement of fact. The purpose is a statement of intent ('to do something'). The purpose should lead to the aim, related to the four parts of the information value chain:

1   'The aim of this report is to identify alternative locations for the head office and assemble the costs associated with each.'
2   'The aim of this report is to define the supplier quality-problem and evaluate the options for solving it.'
3   'The aim of this report is to identify why morale is low, and recommend how to improve it.'

Note the form of words. The aim is very explicit. It should be stated in the report or presentation. It tells the reader or listener exactly what is intended.

The scope of the management communication should follow from the aim. It will help you to structure the communication effectively. The scope may be to

● provide data (facts), without analysis or comment
● provide information (data plus analysis) relevant to the problem in hand
● provide relevant knowledge from a wider, or related, context
● recommend a decision, based on that information and/or knowledge
● say how the related action can be initiated, and by whom

What would be the scope of the management reports in our three examples? For the first it would be to provide data and a small amount of analysis (calculating the costs for each option). For the second there would be more analysis, to identify viable options. For the third there would be a recommended decision and detail of how that decision could be put into effect. The third report therefore has the greatest scope. It might be submitted to a senior manager. If they like it and agree the conclusions they can implement it very easily, by saying 'Yes, do it'.

For a major management communication you should agree in advance with the recipients what the purpose, aim and scope of the communication is to be. If you don't do this, there are several dangers:

- You will be unsure what you are trying to achieve, and will be unable to prepare a coherent communication.
- The recipients may not get what they expect or need. This can lead to frustration and anger. At the least it will waste everybody's time and effort.

Think back to the model of the communication process in Figure 2.4. You are defining the shared context between you and the recipients of your communication. The purpose, aim and scope should be mutually agreed, and stated in the communication itself. They define the framework for what you will write or say.

2.5 Summary

If this summary seems unfamiliar, it is a good idea to go back and refresh your mind about the appropriate part of the details since the ideas provide a framework for the rest of the book.

Section 2.2 looked at the general nature of the process of communication between people. It is concerned with the transfer or sharing of information which is of human significance in the particular situation. There are two theoretical approaches: the semiotic model (of signs) applies in static or slowly changing situations; the process model applies in the dynamic situation of continuous information transfer. These models came together in Figure 2.1, and considered a wide range of types of human communication. It is important to remember that communication involves human perception and our senses can deceive us.

In business, there are four main modes of handling information: text, data, voice and image. Communication is most effective when we use combinations of these modes, such as text and graphics. To convey emotion or conviction in a verbal encounter, words are the least important part of the process.

Figure 2.4 is a model of the communication process. This includes transmission and feedback within a shared business context. Figure 2.5 identified three levels at which one should judge a business communication: technical, semantic and effectiveness. These are increasingly broad in their scope.

Finally, Section 2.2 looked at some practical examples of business communication, and identified some of the limitations on human information processing. People can only absorb information at a rate equivalent to 60 words per minute of English text. However, we normally speak at 150 wpm and read at 240 wpm, as shown in Figure 2.6. To be effective, all business communication must therefore be structured and have summaries. When reading one must adjust average speed, and technique, depending how much of the detail one needs to absorb.

Section 2.3 looked at the role of the business enterprise. It considered the customer oriented view, shown in Figure 2.7. The product on offer always has a service aspect; it may also include a tangible good. The customer will look at the cost, quality and time characteristics of the product and compare them with those of competing products. In international business we are faced with further barriers of time, distance and complexity. From these trends the implications for business communications, both inside and outside the firm, were identified. Technology can have a vital role in the necessary information, manufacturing and logistics systems. The aim is to have a closed loop with the customers, to meet their immediate and evolving needs.

Section 2.4 came in closer to the business itself. It looked at the role of managers and how they use information to support that role. There were two key ideas: the management learning cycle and the information value chain. The learning cycle (Figure 2.8) shows the repeated sequence do–watch–think–try. This is underpinned by the information value chain: data–information–decision–action. In terms of the volume of information the value chain has two major parts: it converges from data to decision, and then diverges from decision to action. It is the agent of change from an old situation to a new situation, as shown in Figure 2.9.

The information value chain provides a model for planning major business communications such as reports or presentations. One needs to agree with the sponsors or recipients what is the purpose, aim and scope of the communication. The purpose defines the context, or reason for making the communication; the aim defines the precise contribution to be made within that context; the scope defines how far along the information value chain it will go. When, eventually, the structure of the business communication is decided this completes the acronym PASS: purpose, aim, scope and structure.

The agenda for this chapter was ambitious: the nature of the communication process, the general role of business, and the methods of management within it. However, the necessary foundation for the rest of the book is now set. As shown in the map of the structure of the book (Figure 1.2), you can now go to any of the detailed chapters (Chapters 3–7). Unless you have a pressing need to jump ahead you will find the numerical order of the chapters the best to follow. I hope you enjoy putting the flesh on the bones already established.

Chapter 3

Saying it

A word fitly spoken is like apples of gold in pictures of silver
Proverbs 25:11

3.1 Introduction

Just to remind you, it is best to read Chapter 2 before studying any of the five detailed chapters. This was shown in the map of the book's structure, Figure 1.2.

Spoken communication, the most natural and common method of human communication, is a highly-complex technique which, in terms of evolution, has set us apart from the rest of the animal kingdom. Speech enables a group of people to achieve concerted action. This is important socially, nationally and in business.

In the business context, we need to do more than simply carry across the style of talking which is adequate in everyday life. We probably need to be more precise and careful. We may have to deal with strangers, including those from different cultural backgrounds. All this presents a special challenge.

This chapter first looks at spoken communication inside the business enterprise, then at communication outside, with the varied people we encounter in the business environment. Finally we look at language itself, and how to avoid some of the possible pitfalls in using it.

3.2 Internal communication

Chapter 1 noted the tendency of engineers to be precise and factual, of nonengineers to use undefined or unquantifiable terms, and of specialists to use jargon (except ourselves, of course). Also how trends in modern management mean that we all have to work together much more effectively as a business team, responding to the changing needs of our customers.

Within the business enterprise this must become part of our corporate culture. That term 'corporate culture' is difficult to define. It has been called the glue that holds the organisation together, what achieves cohesion among the members of the organisation, or just 'the way we do things around here'. Most modern businesses write a vision or a mission statement about themselves. Although a lot of thought goes into preparing these (or should do), the results tend to be anodyne. The statements often refer to organisations which nurture partnerships with their customers and other stakeholders, are caring towards their own employees, and have a developed responsibility towards society and the natural environment. As always, it is not the words that matter, it is whether people collectively 'own' the thoughts and ideas which lie behind the words. That is the difficult bit about changing corporate culture: communicating it and getting people to live, breath and behave it.

Whatever the corporate culture is, or aspires to be, professionals today should choose to act according to those modern business precepts. Here are some of the implications of that:

● All the other people in our business are colleagues. We can't just confine our attention to the one part of the business to which we are formally affiliated. All have to work together to achieve the objectives of the business as a whole.

● We must therefore be able to communicate with all sorts and conditions of people within the business. Sometimes we have to be responsive to the requests they make of us; at other times we expect them to be responsive to the request we make of them. This is the supplier–customer relationship at work throughout the business itself.

How do we communicate effectively with all these different kinds of people: general managers, specialists in our own discipline, specialists in other disciplines? There are some general principles that apply to any mode of communication. For spoken communication they are particularly important because it is in that direct and immediate situation that things can go wrong the fastest. Applying the model given in Figure 2.4, we need to

— Think carefully about the shared context we have with those with whom we are communicating. This means that we won't use jargon with those who are not familiar with it. It means we will pitch our message at a level which is appropriate to the needs and interests of the recipients. We will watch their reaction as we are speaking, and be alert for any signs of incomprehension or even boredom.

— Remember that communication involves a receiver as well as a sender. For spoken communication it is as important to listen well as to speak well. We therefore need to cultivate the skill of what is sometimes called 'active listening'.

— Remember that communication is a two-way process. It involves feedback as well as transmission. We must be alert to the conventions for agreeing to change the direction of communication. Thus in the one-to-one/few situation (conversation) we must watch for the cue which indicates we should change the turn in the conversation. In the one-to-many (presentation) situation we should look for the signs of someone bursting to say something or ask a question. At a higher level, we need to be sensitive to the nature of the interaction which is taking place. This is where transactional analysis (TA) can help.

Look at active listening and transactional analysis more closely. They are techniques that can make our spoken business communication much more effective.

Active listening is about trying to help the other person (the speaker) to communicate effectively with you. First you have to create the right empathy: eye contact without staring; body language that is attentive but not intrusive or threatening; appear relaxed but concerned to hear what they have to say; smile and encourage, especially if they are hesitant. Try to use a venue appropriate to the topic of the discussion. It might be appropriate to discuss an industrial production problem on the shop floor, where the implications can be seen and shared. That would not be the right venue to discuss a pay rise with any of the same people. In an office, do not confront one another on opposite sides of a desk, unless it is a formal or disciplinary matter. It is better to sit around an office table which has no implications of status, or around a coffee table in easy chairs. If there are just two of you, try to sit on adjacent sides of the table, so that you are not staring at each other across an expanse of polished wood. However, ensure that there is a surface for documents, or to take notes, where that is appropriate.

These factors of the venue are important. But there is more to active listening than that. You need to be ACE:

A *Accept the speaker as a person:* They may be very different from you, both as an individual and in terms of their education, job and experience. They may have done something which you think is silly, or wrong. Try to set your personal prejudice apart. Certainly ensure that it is not conveyed, even obliquely, to the speaker.

C *Concentrate on what they are saying:* Set aside the nagging thoughts about all the things you have to do next. Focus your mind on what they are saying. If you don't understand, gently seek clarification in a neutral way at this stage.

E *Encourage the speaker to continue to deliver the message:* Often the subject that you really need to concentrate on is not the one which was the original reason for starting the discussion. It may take some well directed questions to get the speaker to reveal the real problem. This is not necessarily because they are trying to deceive you; it is probably because this is the best process by which, together, you will get to the root of the problem.

In a brief or casual interaction, active listening is little more than giving full and polite attention. In more complex situations it is an important ingredient in the success of a dialogue. Perhaps the most extreme form of this is personal counselling. There, the speaker ('the client') is seeking help from the listener ('the counsellor'). As managers we all have to act as counsellors sometimes, ill-equipped though we may feel for that role. Counselling is different from giving professional advice as a consultant. In counselling, the discussion is centred on the client as a person, rather than on a physical situation or problem. The client must be encouraged to find their own solution to their own problem. This means that the client should do most of the talking; active listening by the counsellor is therefore extremely important. A good relationship must be established, based on trust and the assurance of confidentiality. The counsellor must at no stage be judgemental, even when the core problem emerges. Such core problems usually centre on human relationships. You might despise what the client has done, but the A of ACE means you will not show this.

Active listening is the recipient's contribution to the success of a verbal interaction. Sometimes this is simple. Sometimes, as with counselling situations, it is complex and highly demanding. In all cases, though, it is active rather than merely passive.

One of the aims of active listening is to achieve an adult style of dialogue. Both parties to the discussion are dealing calmly and carefully with the matter in hand. There is no immediate anxiety clouding their judgement. This links to the ideas of transactional analysis (TA), as expounded by Eric Berne (1966 and 1974). Put simply, TA considers three aspects of our character and behaviour, called ego-states. These aspects are called parent, adult and child:

- Parents have standards to which they believe they and others should aspire. They may therefore be critical, angry and judgemental. They may also be nurturing.
- Adults have knowledge and skills. They are analytical, rational and nonemotional. They are concerned with the here-and-now of situations rather than the distant future. They are concerned with learning and change.
- Children may be natural: free and malleable. Or they may be conditioned or adapted to certain views or behaviour. When conditioned, they may be compliant (seeking to please, agree, apologise and be dependent); or they may be rebellious (stubborn, sulky and withdrawn).

These are not three different sorts of people. They are three different types of behaviour which we all exhibit on different occasions and in various circumstances. One way of summing up the difference is to say that the child feels (emotion); the adult thinks; the parent believes. They are aspects of all of us: body, mind and spirit, if you like.

The value of transactional analysis is that it gives us a framework for analysing and understanding the interaction between human beings. It is particularly useful for understanding face-to-face conversation, where the to-and-fro interaction is quickest and most natural. Figure 3.1 shows the model. We use the model by

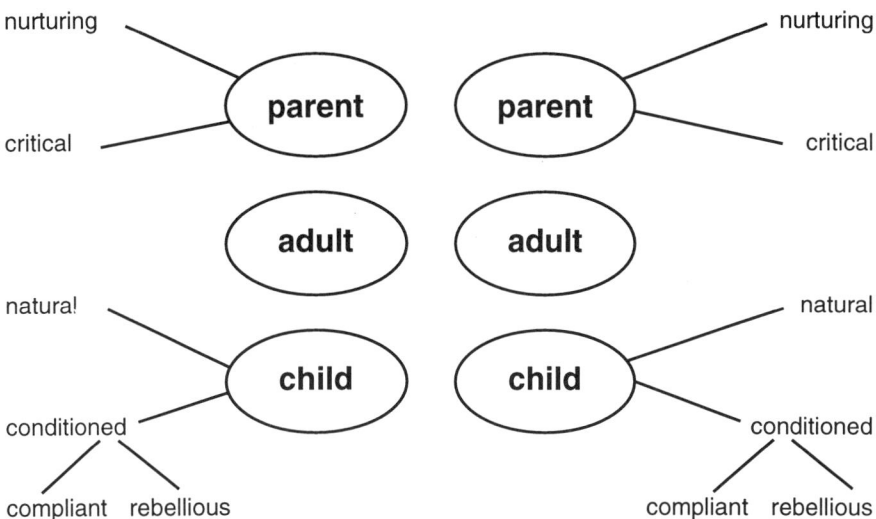

Figure 3.1   Transactional analysis

deciding in which mode each of the two people is behaving, and the nature of their interactions. You can probably think of many examples from your own experience. Try now, before moving on. All the possibilities can occur in the world of business, as well as in personal life. Here are some examples:

1   First speaker: 'I think there might be a better way of doing that job; let me show you how'. This is the nurturing parent speaking. Second speaker: 'Right, thanks. I'm in a bit of mess and really need some help'; this is the natural child replying. There is no conflict, because the nature of the relationship (for this interaction) is accepted by both parties. The two statements are shown by arrows in the first diagram of Figure 3.2. They comprise a stimulus A and a response B.

2   In the foregoing example, the second person might have replied, 'Who are you to tell me what to do? I'm senior to you, and been around a bit in this firm, you know.' This is the critical parent rebuking the child. The diagram for this example then shows the two arrows cross. Both parties are trying to adopt the role of parent and put the other in the role of child. The first person is doing it constructively (nurturing); the second is doing it adversarially (critically). There is likely to be conflict. Perhaps one of them should back off. Probably it would have to be the first speaker.

3   First speaker: 'Hello, Jim. Could we have a word about the marketing of the new widgets? I think we may need to arrange some more training for the salespeople'. This is the adult, talking rationally about a particular business situation. If Jim replies, 'Yes, that could be so. Shall we meet tomorrow at four? I'll ask Jane to come as well', this is the adult responding constructively and rationally. The third diagram in Figure 3.2 shows this adult-adult interaction.

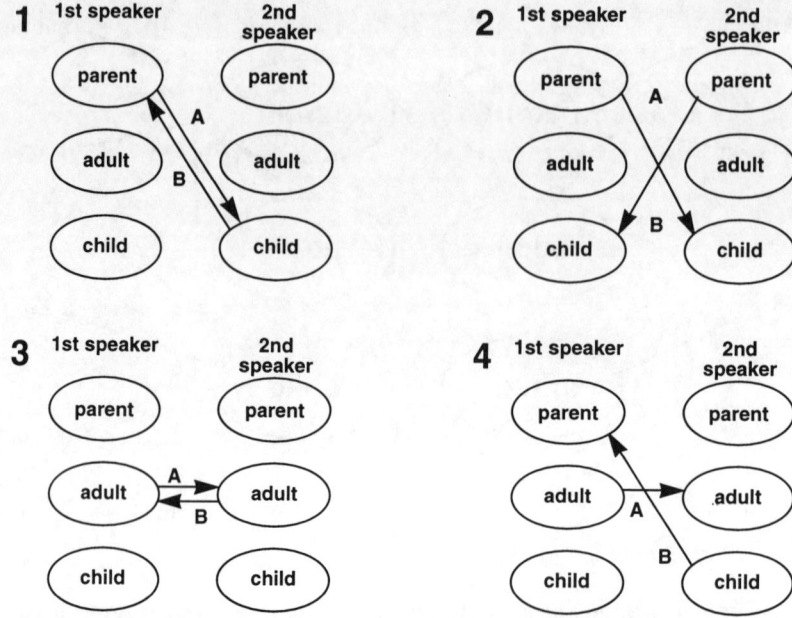

Figure 3.2　Analysis of dialogue

4　Jim might instead have replied: 'Yes, of course, if you say so. I know the figures aren't too good at the moment. We need your advice, and we'll do whatever you think is necessary'. This would be the compliant child seeking a subservient role to the parent. From a management point of view it would be pretty feeble. In any event, the two arrows cross and there is a potential problem.

These examples are perhaps extreme. However, they do show that problems are likely to occur when the stimulus and response arrows cross on the diagram. In most social situations we try to achieve adult–adult interactions. An exception of course would be people dealing with their own young children. There the adult–child interaction of transactional analysis would be appropriate for much of the time. Unfortunately some people don't recognise when their children reach the age when they begin to need adult–adult interactions. It's sooner, and more often, than we might think. Children should be treated as adults whenever possible, and there is a similar message for business.

　In nearly all business situations we should aim to achieve adult–adult inter-actions. The exceptional cases are where people may need disciplining, after normal management measures have been tried and failed, or where they are seeking counselling. Here are some of the things that can go wrong in the normal business situation, making an adult–adult interaction difficult to achieve or sustain:

● The first speaker adopts the position of a critical parent. This could be by being pompous, 'putting people in their place', or by flaunting rank or status.

- The first speaker makes an adult stimulus, but the second makes a subservient response (child to parent). This is the problem of people who will not express their views, are overconscious of their lack of status, or are not assertive enough. This is as bad as the first case for preventing an effective adult–adult dialogue.

Transactional analysis is a useful model for understanding dialogue. Remember that the dialogue will include both verbal and nonverbal communication. It helps you recognise the roles which people adopt temporarily for individual interactions. It helps you recognise the dangerous situations where a few words can upset what was a stable and constructive dialogue. Think about how the model applies in situations you have recently been involved in, at home or at work.

So much for active listening and transactional analysis. They help the communication in two important ways. Let us now turn to the main active role in verbal communication: actually saying it. How do we accomplish this task?

Well, start by drawing an important distinction. This is between what we do (the task) and how we do it (the process). This applies whether we are working alone or in a group, and is shown in more detail in Figure 3.3. The task is what

TASK	PROCESS
what we do	how we do it
formal agenda	informal agenda
task motivation	personal motivation
hierarchy	pecking order
rules	group norms
facts and figures	feelings and emotions
'THE WORDS'	'THE MUSIC'

Figure 3.3   Task and process

we have been asked to do, or set ourselves to do. There are usually specific outputs and they must be achieved in a specified time. The task can often be viewed as a small project, which we must manage accordingly. Each of the people involved has a clear role, in terms of their job or responsibilities in the organisation. The hierarchy is clear; everyone knows who is the boss and what the organisation chart looks like. We may be governed by rules set out by the organisation or the wider world (laws). We are dealing with facts and figures. The task is objective and unemotional.

The process is not so clearly defined, but it is equally important. The people involved may all have their private agendas. They may be anxious to please the boss, wishing to do well on this particular job, or settle a score with an old rival. People will establish a pecking order of mutual respect which may be different from the formal hierarchy. They have, or may develop, their own procedures for working together to achieve the task. They have feelings about what is going on: commitment, comradeship, trust, and friendship. Or it might be the reverse of all these. In any event these factors will greatly affect what goes on (the process) when the people try to work together to achieve the task.

The distinction between task and process can be summed up as 'The Words' and 'The Music'. Both are happening at the same time, as they do in opera. You can only give your full attention to one of them at a time. But you must attend to both, alternately. In a group of two or more people, if the process starts to go wrong anybody should be able to stop the work on the task and ask that they review the process before things get out of control. Using the metaphor, if the music gets too loud you can't hear the words.

Most engineers and other technical people are highly task oriented. They are practical people, concerned with getting things done. They don't want to spend time thinking about procedures, human relationships in groups, team roles and all that. They want to get the job (the task) done. There is nothing wrong with that aim, but it can be seriously frustrated if the process goes wrong. If people don't work together as a team, can't reach agreement, or refuse to shift their positions you have to do something about it. Otherwise the task will not be achieved, or at least will not be achieved effectively.

Whenever people meet together in a business situation the task and the process are both there. As a manager you need to be conscious of both, and exert appropriate influence for both. Think what that involves for a verbal dialogue.

For the task of speaking it includes

- Dealing with the facts of the situation: who, what, when and where.
- Avoiding jargon, and starting with the shared context which you have with the other person.
- Following the ABC of business communication: accuracy, brevity and clarity. Make sure the facts are right (check them if necessary); be as concise as possible, avoiding irrelevancies; structure what you say so that it is as clear as possible.

All that is about the task of conveying the information effectively. What about the process: the how and why of the situation? Here are some suggestions:

- Establish an appropriate venue and physical conditions for the dialogue. This was discussed earlier in this section.
- Use the transactional analysis (TA) model to understand the interaction. As a speaker you should be in the adult mode nearly all the time.
- Watch for signs that the listener may be losing interest, not understanding you, or trying to say something in reply. Be ready to change the turn for speaking.
- Check that the listener has received and understood your message. Recall Figure 2.5 in Chapter 2. You must be effective at the three levels: technical, semantic and effectiveness. Gently, you may have to ask the listener to repeat the key points to you, just to make sure.

- Don't be purely mechanical; add some music to the words. Try to put people at ease in your initial greeting; chat informally for a few moments before you start into the serious business matter.
- Similarly at the end, when the formal business has been done, round off the dialogue with a few pleasantries. Ideally you should both part with the feeling that you have had a pleasant (as well as a fruitful) experience. Don't spend a lot of time on idle gossip or chatter, but equally don't get the reputation of being humourless, or unable to talk about anything except work.

3.3 External communication

So much for internal communication. There, you have one important advantage in seeking effective verbal communication: you know each other, or at least you are colleagues within the same business. You share the same corporate culture and can get to know each other fairly easily.

With external communication it is different. You cannot assume that everyone is on your side, or even the firm's side. You are dealing with people whose legitimate interests differ from yours, notably the customer. You share with the customer an interest in getting the right job done effectively. But when it comes to issues of cost and time, your perspectives may be very different. And who is my customer? The management trends of Total Quality and market orientation mean that we all have to deal with customers; some will be internal to the company and others will be external. In each case the supplier–customer relationship makes the process of business communication different.

So, for external communication you may be dealing with strangers. Some of them may be from different countries and cultures than your own. You have to apply all the principles outlined so far, plus some extra ones as well. This section considers several aspects of that: negotiation, interviews, cultural differences, gesture, the media and making speeches.

In a negotiation situation the proceedings are necessarily quite formal. Everything that each person says is noted by the other party and can influence the development of the negotiation. You need to plan your course of action carefully. Here are some specific points to consider:

- In the planning phase, before the formal negotiation starts, you need to consider your own objectives and criteria. What would you ideally like to get from the agreement? What is it absolutely essential for you to get? Consider the time and cost aspects of this, too.
- Do a similar appraisal for the other party. Mentally, put yourself into their situation, and try to answer the same questions.
- Compare the two positions. Are they compatible, or mutually exclusive? In the first case it should be relatively easy to reach agreement; in the latter case it will be impossible unless one or both parties is prepared to shift their position. Consider how much you are prepared to give, in the interests of achieving

agreement. You should concede things which are most attractive to the other party but which cause the least possible pain and anguish to you. If you expect to have to concede points, it's best to start from a high position.

— When it comes to the actual face-to-face negotiation, the points mentioned earlier for internal communication are very important. The venue, timing and room layout can all have a significant effect. Most people are at ease and most confident on their home ground, but clearly both parties cannot have that advantage (unless they are using telecommunications, as discussed in Chapter 7).

— The style of your dialogue will depend on what you are both trying to achieve. You may be trying to settle a dispute, ratify an agreement, secure an order, or appease a dissatisfied customer. In each case the style will be different. You must pay full attention to body language as well as the words you utter. Listen carefully; take notes; use active listening to tease out information in a difficult situation.

— Round off the negotiation in a civilised manner. If you have reached agreement this is easy; social chat, and perhaps a meal, may follow. If you have not reached agreement, it is more difficult. Try to sum up objectively and identify the particular issues on which you have failed to agree. This may be the starting point for a further session of negotiation. You may need to fix a time and place for that.

— In any event, try to avoid the dialogue getting heated or angry. If emotions get high, try to defuse the situation by shifting temporarily to some noncontentious or purely factual aspect of the situation. When people have cooled off, return to the contentious issues calmly.

It is not appropriate to discuss the detail of negotiation skills here. For further information, see, for example, Fowler (1990). A special type of negotiation situation is the interview, especially when this relates to a job application; see, for example, Peel (1990). But we can agree that external communication, especially negotiation, is not easy. It involves people with different viewpoints and opposing criteria of success. These difficulties can be greater when the communication is between people from different cultures, whether national or international.

National cultural differences can be quite marked. Even in the UK, there may be noticeable differences of business or social behaviour between people from England, Wales, Scotland and Northern Ireland. Within each of those countries there can be marked differences, for example between the south-east, west and north of England. However, it must be said that these differences are becoming less marked, not least because of the influence of business conducted over greater distances.

So it is international cultural differences which we most need to attend to. They affect both the task and the process of what we do. Take, for example, the business of project management. A project is a piece of work which must be carried out to a required standard in a specified time. It is one-off rather than repetitive. Surveys have shown that different national groupings are on the whole better at different stages of a project. Turner (1993) reviews this by reference to the work of Hofstede, who examined national differences in terms of four parameters:

Power distance, P: This is the extent to which a less powerful person in a society accepts inequality of power, and considers it as normal.

Individualism, I: This is the extent to which individuals primarily look after their own interests and the interests of their immediate family (spouse and children).

Masculinity, M: This is the extent to which the biological difference between the sexes is used to define different roles for men and women.

Uncertainty avoidance, U: This is the extent to which people are nervous of situations they consider to be unstructured, unpredictable, or unclear, and the extent to which they try to avoid such situations by adopting strict codes of behaviour and a belief in absolute truths.

Hofstede's studies revealed two strong groupings: people in developing countries tend to have high-P and low-I; people in developed countries tend to have low-P and high-I. Thus the individual has a greater prominence in developed countries. The factors M and U, although varying widely between individual countries, do not reveal a broad grouping similar to that for P and I.

Turner (1993) reviews the evidence in respect of the three major phases of project management: initiation; planning and execution; and closeout. He finds that project management is typically a Western approach, with Germany being most suited to it. Arab countries and east Africa perform well, but developing countries score rather low. Thus, when engaged in a project with an overseas partner you need to take account not only of the nationalities involved but also the stage of the project. There are also general differences of management style. Some of these are anecdotal, but others are built up from hard-won experience. When starting to conduct business with a new country, you should therefore seek advice and study the business culture more closely.

And so to the general social differences between national cultures. These are especially important when speaking to people face to face. You need to know the appropriate conventions of nonverbal communication: body language and gestures.

Morris (1994) gives a fascinating view of the human species, including the way we use gestures. Each of us can use something like 3,000 different gestures with our hands and fingers alone. Some of them merely add light emphasis to what we are saying, and may pass almost unnoticed. Others are very strong in their meaning, and may override any words which are being spoken at the same time. The gesture becomes the message, especially if it is rude. We therefore need to be specially careful to know not only what is polite conduct in a particular country but also what might be construed as rude. Even within Europe there can be marked variations:

● Touching the earlobe in Portugal indicates that something is especially good; in Italy it is used to accuse a man of effeminacy; in Spain it denotes a sponger who never pays for a drink.

- In most countries the V-sign made with the first two fingers and the palm towards the face is recognised as a sign of victory. In the UK it is a very rude insult.
- Making a ring of the thumb and first finger means OK in the USA. In Sardinia and the middle east, it is obscene. In Japan it means money.

So you have to be careful to avoid giving offence. Equally, though, there are many widely accepted gestures which can give emphasis to a speech or business presentation. Morris (1994) lists the pinch, the punch, the chop, the embrace, the prod, the pat, and the semiclenched fist. For greater detail about gestures, see Argyle (1988) and Mead (1990). For business aspects of international cultural differences, see Silk (1995).

There is a special type of business communication which often has a cross-cultural aspect. This is dealing with the media - television, radio and the other news vehicles. In recent years there has been a trend towards a more frank and direct style of interviewing and reporting. Most people welcome that, but interviewers show occasional excess of zeal. They can be overbearing or aggressive. Words can be reported selectively. You may not be given adequate time to respond to a question before the interviewer asks the next, or interrupts you in some other way. Dealing with the media requires a cool head and a clear mind:

— Think out the message which you want to convey well before the interview. On an important topic you may need to get formal agreement or clearance from within your organisation.
— Make sure that you are not rushed before arriving at the venue. If possible have the interview on your own territory. Make sure you dress appropriately.
— Tell the interviewer, before you start, whether your remarks are 'on the record' or 'off the record'. If they are on the record, they can be published and attributed to you by name. If they are off the record, the interviewer can use your remarks only for background information or report it with an oblique attribution which does not enable you to be identified personally. There are tactics to consider here.
— During the interview, be calm and rational. If you get heated, you will convey a bad impression about yourself and undermine the cogency of what you are trying to say.
— Allow the interviewer to take the initiative, because that is their job. If they have done their homework properly you will be asked the key questions about the topic under discussion. Make sure that in responding to those particular questions you convey the planned message which you view as important. Say it slowly and with conviction. This makes it stand out from what is often a frenetic pace of media reporting.
— Be robust in preventing interruption when you are answering the key questions. Most interviewers will respect your request to be allowed to complete your answer.
— If you are not asked the key questions then take the initiative yourself. 'Yes, but I think the key issue is really this. . .'

Dealing with the media is an important aspect of corporate communications. For more information see Fiske (1982), Ind (1992), Janner (1988), and McCall and Cousins (1990).

Media interviews can be stressful. There is another type of external business communication which can be equally stressful, but in a different way. That is speech making. By that I do not mean a formal business presentation; that is dealt with in Chapter 5, and should certainly not be a speech! No, here I mean the speech made at a social occasion such as a dinner.

The after-dinner speech can be a worry for many people. As with any other challenge, the key to success is in careful planning. Decide beforehand exactly what you want to say, and the balance between serious and light-hearted style. The occasion often dictates what you must say; this might be to thank a colleague on retirement, welcome a customer, or mark an anniversary. But the rest is very much up to you. Here are a few tips:

- Until you gain experience of public speaking, keep it short and simple: directly related to the occasion and what needs to be said. Make sure you know who else will be speaking, and in what order. Ask about the procedure for initiating the speech.
- Try to enliven what you say with with relevant anecdote or a story which is directly relevant to the topic in hand. However, do not use the occasion to embarrass anyone present.
- Make sure you have complete silence before you start speaking.
- Be very careful about humour. Some people can remember jokes and some are good at telling them. Others can do neither. You know your own capabilities in this respect. Play it safe, and make sure that humour will not cause offence. The safest, and most gracious, is humour at your own expense. Never laugh at your subject, or unkindly at other people. Dry humour is a British speciality; it adds lightness of touch, and does not fall flat if no-one laughs out loud.
- Most libraries have books of after-dinner stories and jokes. It is useful (and enjoyable) to look at some of these, and make a note of stories or jokes which amuse you and might be useful. It is best, though, to adapt the context of a story to fit the occasion. Then, even if people have heard the joke before, they will be amused by seeing it applied to a different situation.
- For planning your speech and using notes, use the advice in Chapter 5. More general advice is in Janner (1988).

3.4 Using language

Now focus on the primary channel of communication: language. Language has developed over thousands of years to help human beings co-ordinate what they think and do. Section 2.2 showed that the speed at which we speak and listen is far greater than the speed at which we can digest and understand new information. Language must therefore be used so that the main message is repeated and reinforced. In social conversation we do this almost without thinking. In business communication we have to be a bit more careful. Here are some issues which arise:

— A common accusation against scientists and technical people is that they are fairly normal in their social conversation, but as soon as they start talking about their work they become obscure and unintelligible to the nonspecialist. It's as if their training made them go into a different mode of thinking and communication. This could be a result of being required to answer examination questions, write technical reports or scientific papers. Surely we should aim to communicate verbally with the same ease and fluency which we would aim for in everyday dealings?

— Being too concise so that every word carries important meaning is dangerous. It is contrary to the natural mode of using language, and can lead to disaster. For example, the pilot of a Boeing 747 at Tenerife airport said to the air-traffic controller 'We are now at takeoff'. He meant 'We are in the process of taking off', but the controller thought he meant 'We are waiting at the takeoff point'. As a result of this misunderstanding two aircraft collided, with the loss of 583 lives. For this type of specialist communication clear conventions need to be established, then learnt and adhered to.

— Everyone knows the game of Chinese whispers, where a message is whispered privately from one person to the next. Sooner or later the message becomes seriously distorted. In a similar way, gossip on the informal communication network (grapevine) becomes distorted, sometimes deliberately and sometimes innocently. Rumour is a great enemy of management; frank and timely corporate communication is the antidote.

— In both social and business communication we must be careful not to cause offence by our choice of language. Political correctness is a recent movement which highlights the issues involved. To avoid controversy, we must not make implicit judgements about gender roles, disability or other aspects which are not relevant to what we have to say. For example, by referring to a manager as 'he' you may be implying that most managers are men. That may be true, but it may not be just. By using the plural 'they' you can avoid this problem more elegantly than by saying 'he or she'.

Within the natural constraints of language, we must aim to convey information at a rate which people can deal with. Accuracy, brevity and clarity (ABC) should be the features of the core message. Repetition and reinforcement should make the core message memorable. We should avoid ambiguity and emotion. We should aim for logical persuasion, rather than rhetoric. In short: be logical where you can; be persuasive where you can't.

Logical thinking, leading to logical speaking, is therefore a fundamental skill of spoken business communication. It will help to look at some examples of the problems of logical thinking expressed in language:

Using language ambiguously: It is difficult to use language without any chance of ambiguity. To do so is often so cumbrous that we lose the main message.

That is one reason why mathematical and logical notations were developed to express ideas in those fields without risk of misunderstanding. However, we can and should take care to use language to avoid obvious doubt:

- 'He noticed her shaking hands'. Was she greeting another person with a handshake, or were her hands trembling?
- 'She has been at the bar a long time'. Has she been drinking for several hours, or is she a lawyer who has been a barrister for many years? The second interpretation will only occur to someone who knows about lawyers and the bar; many countries have a legal system without that terminology.
- 'Let me know if you want to borrow the book'. Does this mean I have to contact the speaker, even if I decide not to borrow the book? It might be better to say 'If you want to borrow the book, let me know'. Then I know that I don't have to say anything if I decide not to borrow the book.
- 'Be logical where you can; be persuasive where you can't'. Yes, that's the motto just suggested! It is intended to be short and balanced, to make it more memorable. It is an aphorism. Yet the second part is paradoxical; how can you be persuasive where you can't? The key, of course is that there are some words which are implicit but which have been omitted. 'Be logical where you can; be persuasive where you cannot be logical' is less ambiguous but also less memorable. The little paradox evident in the shorter version actually helps to remember it.

Self-interest: If the outcome of a business discussion will have some direct effect on us personally then we must be specially careful. It is easy to have a conscious or subconscious bias in what we say or do. We must be careful to be professional and objective. To make this clear, it is usually best to declare a personal interest, especially if it is financial. In extreme cases we may feel that it is not appropriate to deal with the matter, and we should seek to be relieved of the task. Declaring a personal interest may be important in a brief discussion, a meeting or in a written report. It avoids the danger of bias and any chance of retrospective suspicion or accusation.

Habit: We all make assumptions and get into habits. We have to, to survive practical everyday life. But habits of thinking can be very dangerous. They can close our minds to new or different ways of doing things. That may be comfortable, giving us what has been called the illusory confidence of a closed mind. However, we should really try to think of new options ourselves. Equally, we must be receptive to the suggestions of others. The tendency to reject without careful thought any suggestion coming from another person, department or organisation is called the not-invented-here syndrome. It should have no place in modern business. Ideas should be assessed on their merits, irrespective of who happened to think of them. If you find you have a tendency to reject ideas out of hand, make sure you think more before you comment. Bite your tongue, or count to ten, or ask other people what they think first. Don't just give a negative reaction right away.

Emotion: Emotion and rhetoric seek to influence people without taking the trouble to persuade them. For the core of your argument, don't rely on emotion. No logical argument can be based upon unsubstantiated emotional assertions.

Suggestion: This is almost as bad as pure emotion. It suggests an idea, while seeming to make it plausible and acceptable. A lot of advertising is based on the fact that we are open to suggestion. Avoid phrases like 'We all know that. . .' and 'Clearly. . .' (when it is not at all clear, logically). They suggest to the listener that they are a bit odd, or dim, if they don't agree with the view you are suggesting. Such phrases can be used unscrupulously, to mask a lack of evidence.

Misuse of authority: Some eminent person may be quoted to add authority to statements or views which are not supported by other evidence. This is fine if the eminent person is an expert speaking within their own acknowledged field of competence. Otherwise, it is a misuse of authority. Therefore avoid phrases like 'The managing director feels we should. . .'. The MD may only have expressed an informal or provisional view. It is different if the MD has made a clear decision, which is an assumption for you to take and build on. Thus you should quote authority in two circumstances:

- administrative authority, when a clear 'order' has been given.
- expert authority, when the expert has appropriate qualifications or acknowledged expertise in the relevant field, and you are prepared to trust their advice and judgement

Just as we should not misuse authority, so we should not ignore ideas which come from humble sources. Lord Melbourne (1779–1848) said 'Never disregard a book because the author of it is a foolish fellow'. That remark is a consolation to authors everywhere.

Avoid errors of logic.

This last item deserves closer attention. To do that, it helps to be aware of the two main methods of logical thinking: deductive and inductive. The structure of the simplest form of deductive logic is shown in Figure 3.4.

This form of deductive logic is called the syllogism. It starts with a general rule, sometimes called the major premise. It identifies a particular case which is covered by the general rule. Then it makes a deduction about the particular case, by applying the general rule. Thus deduction argues from the general to the particular. To use deductive logic in this way you must be confident that the general rule is well established, and that the particular case comes within its scope. This is the case for the classical example shown in Figure 3.4.

A common misuse of deductive logic is by omitting to state the general rule (the major premise). The error is called the obscure major premise, and here is an example: 'State control of industry is an infringement of personal liberty, therefore we must resist it'. This has only two of the three parts of a syllogism.

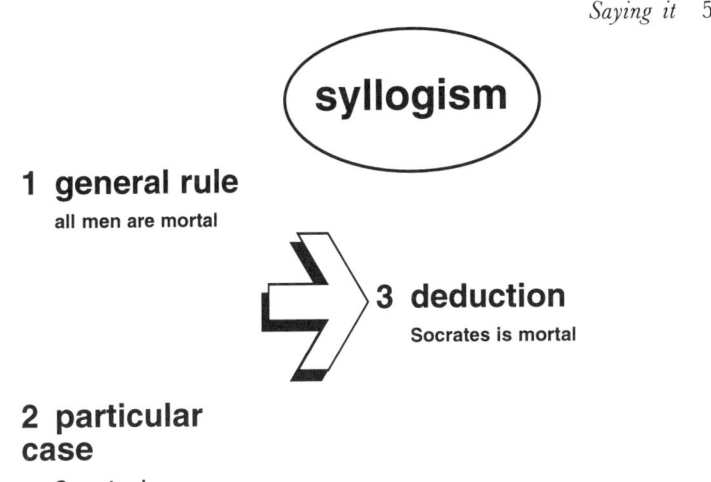

Figure 3.4   Deductive logic

The general rule (which is implied but not stated) is: 'All infringements of personal liberty should be resisted'. Most people would not accept that as a general rule. By omitting it, the speaker makes the argument appear more plausible; it is a form of suggestion.

The second main method of logical argument is induction. Inductive logic draws probable general inferences from a set of particular facts. An example is the statement 'All swans are white'. One might formulate such a rule after seeing a large number of swans, all of which were white. However, that would be wrong because in some places there are black swans. The first time we see a black swan we know our general rule has been disproved.

Again, one can misuse this form of logical argument by generalising from what is clearly inadequate data. For example, the statement 'All Australians are drunkards' may be based only on the fact that the speaker met two Australians last Saturday night and they were both drunk. It is also possible to make innocent mistakes, like saying 'All birds can fly'. Then we remember penguins, ostriches and the like. However, a zoologist can say 'All birds have feathers', because that is the distinguishing feature by which zoologists choose to call something a bird.

Induction appears to be a weaker form of logical argument than deduction. But is this really so? There are weaknesses with both forms. How can one be sure of a general rule for deduction if a new fact could turn up to falsify it? The general rule for the deductive syllogism was often established by inductive reasoning from limited data. Sometimes this is based on so much evidence that there can be little reasonable doubt. Examples of such highly reliable general rules are 'All men are mortal' and 'The sun will rise tomorrow'. Often, though, we have less evidence. So look critically at the underlying evidence for so called general rules, and the logic of what may appear to be a sound argument. There may be a concealed gap in the logic. Even the statesman Sir Winston Churchill is said to have annotated the script of one of his speeches 'Argument weak; speak loudly'. The philosopher Stewart Hampshire paid a great tribute to Bertrand Russell (1872–1970) when he said that Russell would never use rhetoric to fill a gap in an argument.

Students of philosophy will be outraged at this gross simplification of the problem of certainty of knowledge, but it is adequate to highlight the dangers lurking in practical business situations. You might be interested in reading further on this topic. The work of Sir Karl Popper, who died in 1994, is of particular interest for scientists and engineers. He changed the way we think about scientific method, formulating the principle of falsifiability for useful scientific hypotheses. See Popper (1994/A, 1994/B and 1995).

Here are some further examples of errors of logic. Remember that we are not concerned here with the truth of the statements, only with their logical self consistency!

— 'I was drunk after taking three different types of whisky with water. So it must be the water that makes me drunk'. There appears to be some logic to this, but it is a case of false cause/effect. Because something precedes another thing it does not necessarily cause it.
— 'The Bible is true, because it says so' is an example of circular argument. An assertion is made without bringing forward any external evidence. 'This statement is true' is a condensed version of that, but it is at least self consistent. However, 'This statement is false' is not self consistent. It becomes a paradox, like the Zen idea of 'the sound of one hand clapping', or the statement by Heraclitus (who died about 460 BC) that 'All generalisations are false, including this one'.
— 'As using a razor makes it blunt, so using your brain too much will make it blunt'. This is an example of false analogy. Remember that analogy is the weakest form of logical argument. If anything, using the brain will sharpen your mental faculties. Console yourself with that thought as you wrestle with these logical paradoxes!
— 'Jones was not intoxicated today' is suggestive. It suggests that Jones is usually intoxicated.
— 'Your statement that smoking is harmless is untrue, because you are a smoker'. This is an argument against the speaker rather than what they are saying. It implies that the speaker cannot be objective.
— 'Every attempt to prove that man is not immortal has failed. No evidence can be found that men's souls do not exist after death. Hence immortality must be true'. See earlier remarks about Sir Karl Popper and the principle of falsifiability.
— 'Some numerate people are musicians. All engineers are numerate people. Therefore some engineers are musicians.'

Is that logically sound or not? Think about it for a moment. It is by no means obvious, and it is easy to get tangled up with the words. The best way to resolve it may be with a Venn diagram, as shown in Figure 3.5. The circles denote sets of entities. The first statement ('Some numerate people are musicians') means that

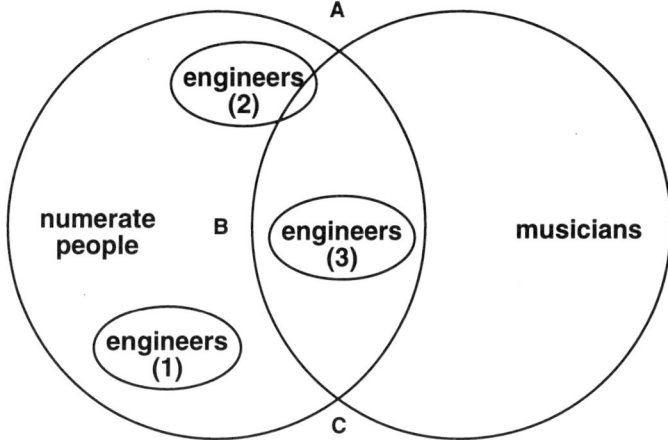

Figure 3.5   Venn diagram

one must draw overlapping circles for the two sets numerate people and musicians. Notice that even in doing that we are making some assumptions which are cannot be strictly justified from the statements:

- there are some numerate people who are not musicians; and
- there are some musicians who are not numerate people.

The second statement ('All engineers are numerate people') means that the set 'engineers' must be lie completely within the set 'numerate people'. However, you now see the problem: we have no way of knowing where the set 'engineers' lies in relation to the boundary between numerate musicians and numerate nonmusicians (arc ABC on the diagram). There are three possibilities, all of which lead to different statements about the musical prowess of engineers. The example actually assumed the case marked 2, but there is no justification for that assumption. The deduction is therefore not logically sound. Note that the argument was not in the form of a syllogism, because the first statement was not a general rule.

So, overall, logic can be a minefield! It is important to be aware of the ways in which arguments can appear convincing but actually be unsound. Nobody expects you to speak with complete logical precision all the time; that would be intolerably boring, and probably impossible. However, you should try to avoid the more obvious pitfalls of ambiguity and logical error when speaking in a business situation. When writing, of course, you should be even more meticulous!

And remember the maxim: be logical where you can; be persuasive where you can't. Here is a thought from the writer Anthony Jay: 'There is only one area of publication in which men proceed by remorselessly logical steps from unquestionable premises to unarguable conclusions, and it is called science. The beauty of science, and in particular mathematics, is that you can convince without having to persuade. But in all other areas, persuasion is necessary'.

3.5 Summary

This chapter started by recalling the challenges identified at the beginning of the book. Section 3.2 discussed internal communication in the modern business enterprise. How we do it will depend partly on the corporate culture: the customs of behaviour for that particular enterprise. Whatever that corporate culture is, today's professional should aspire to the style appropriate to the caring and responsive type of business.

Applying the model of the communication process from Chapter 2, there were three main implications of this. First, one must consider the shared context for the participants in the communication process. This means, for example, using only language and terminology which they all understand. Secondly, one must remember that communication requires listening as well as speaking. The technique of active listening means using the most appropriate venue (including the seating arrangement) and then using the ACE principle: accepting the person speaking; concentrating fully on what they are saying; and encouraging them to proceed. The skills of active listening are needed most in the counselling situation which all managers find themselves in from time to time. The third implication of the model was managing the dialogue which, in verbal communication, can change turn and proceed very quickly. There, the model of transactional analysis was useful. During a particular interaction the two participants may each assume the role of parent, adult or child. In business, the adult-child relationship is necessary just occasionally, but it must be in its positive (stable) form. More usually we seek adult-adult interaction, where the dialogue is rational, calm and professional. The TA model helps us understand the situation, and its potential dangers.

Finally, Section 3.2 considered the actual speaking. As for other business activities, we need to distinguish between the task and the process (the words and the music). The task of speaking centres on using the transmission and feedback parts of our model of human communication. The process aspect means creating the right situation for the dialogue, checking that people understand, and adding some social grace to the formal business of conveying information. Try to make it an enjoyable human experience, rather than a cold and mechanical transaction. Section 3.3 turned attention to external communication. The trends of total quality management and customer orientation mean that we must increasingly deal with people outside our immediate workgroup. They are all our customers, and will therefore have a different perspective of the business situation. They will also have different criteria for the successful outcome of business communication. This is expressed in the art of negotiation, where planning and careful thought about our own position and that of the other person is very important.

All these situations are more complex when dealing with people from another national group or a different business culture. The section looked briefly at national differences in respect of project management; there are marked differences between developing and developed countries. Finally, it looked at social differences, and the importance of gesture. The meaning of a gesture can be totally different in different countries. We must study this if we are to avoid giving offence. Equally, though, there are benign gestures that are widely accepted and which can reinforce the impact of spoken words.

Section 3.4 considered some of the problems of using language in a business situation. They include the ambiguity of language, self-interest, habit of thought (especially the NIH syndrome), emotion, suggestion, misuse of authority, and errors of logic. The section looked in detail at common errors of logic, because they are very dangerous. Both of the main methods of logical reasoning (deduction and induction) are open to abuse. There are other, more subtle errors where the words, or their absence, conceal the gap in the argument. There comes a point where words are not really adequate to resolve the issue; a Venn diagram may then help.

Overall, then, one must remember that speaking is a two-way interpersonal activity, set in a business context. Style will depend on the people we are dealing with and the topic to be discussed. For the task of communication we must have accuracy, brevity and clarity (ABC); that includes being logical. For the process, we should make the exchange as pleasant and fruitful as possible. If you want to consider this topic of verbal communication further, see Hamlin (1988).

Effective speaking is the foundation for all other methods of business communication. The author Quentin Crisp remarked that every time we speak we can make some situation better or we can make it worse. Before speaking, it's worth thinking which of those two effects we shall have.

Consider, also, the quotation given in Figure 3.6. There's a puzzle there: has the word 'not' been omitted at one point?

Thoughts and Expressions

In promulgating your esoteric cogitations, or articulating your superficial sentimentalities and philosophical or psychological observations, beware of platitudinous ponderosity. Let your conversational communications possess a ratified conciseness, a compact comprehensiveness, a coalescent consistency and a concatenated cogency. Eschew all conglomerations of flatulent garrulity and jejune babblements. Be not inebriated in the exuberance of your own verbosity. Let your extemporaneous descantings and unpremeditated expatiations have intelligibility, psittaceous vacuity, ventriloquial verbosity and vaniloquent vapidity.

Figure 3.6   Thoughts and expressions
Handwritten by Frederick Charles Archer (1891–1975)

Chapter 4
Writing it

The weakest ink is better than the strongest memory
Chinese proverb
Polonius: 'What do you read, my lord?' Hamlet: 'Words, words, words'
William Shakespeare (1564–1616)
All writing is a form of prayer
John Keats (1795–1821)

4.1 Introduction

The written word lacks immediacy, but it has the potential to be a much more carefully argued and logical form of business communication. It can convey complex ideas and facts in a way which a continuous spoken message cannot. The most important ideas in business usually have to be set down on paper. The very act of doing so is a discipline which tests and refines the logic of what we wish to say.

Many of the principles discussed in Chapter 2 for spoken communication apply equally here. These include: understanding the shared context of sender and receiver; using appropriate language; avoiding the logical pitfalls. Some of them apply to an even greater extent in the case of written communication. I shall therefore focus more closely on them.

4.2 Written communication in business

In discussing spoken communication, Chapter 3 distinguished between internal and external communication. Internal communication was with immediate

57

colleagues, inside the firm or a particular department within it. External communication was with people outside the firm (notably customers) and with those people inside the firm with whom we may have a fairly formal relationship. This might be, for example, because we have an internal supplier-customer relationship with them. External communications need to be slightly more formal, careful and considered even when they are only spoken.

The same applies, to a greater extent, in the case of written communication. By putting something on paper, or into an electronic messaging system where it will be stored long-term, we are doing something more permanent than uttering words which are immediately lost. Written material can be looked at again, and criticised with the benefits of hindsight. We sign it. This means it is 'on the record', attributable in every respect to us. We take responsibility for it. Our reputation depends on it. We must be more careful, not just to protect ourselves but to make sure the written communication achieves what is intended of it.

Here are some examples of written business communications:

> *The memorandum, or memo:* This is a short communication, usually to one or more colleagues on a particular topic. Speed is usually important, so the sender usually prepares it directly. It is important that the physical mail system achieves a delivery time which matches the urgency of the matter discussed. To achieve this some large organisations have a multilevel priority system, especially if several sites are involved. It is important not to abuse priority systems; that degrades the service for everyone. Priority levels should ideally be defined in terms of the target transit delay. In practice, however, they are usually relative priorities; users then have to use their experience of actual transit delays to decide which level to specify.
>
> *Electronic mail, or e-mail:* This is the IT-based messaging system which is often an adequate substitute for short paper-based written communications. Like the memo, in terms of formality it lies somewhere between the short chat or telephone call (informal) and the written letter or report (formal). The great advantages of e-mail are speed, and the ability to be selective about which (if any) items you need to print out for physical storage. However, most e-mail systems are not very convenient for large documents; it is more difficult to browse or flick through pages on a screen than it is with the paper version. Also, everyone you are communicating with needs to have access to the system. Large modern companies tend to have pervasive office automation (OA) systems which include e-mail and other features, but many smaller companies do not. The special points about using IT-based systems are discussed further in Chapter 7. Some e-mail systems have a priority system for urgent messages, the ability electronically to 'sign' the message (to prove it came from you), and a delivery advice system (to prove the addressee received it and displayed it on their screen). You cannot, of course, prove that they read and understood it; that is the challenge to your writing skills.
>
> *The letter:* This is a longer and more formal method of written communication, usually external. The memo or e-mail are usually written straight off, perhaps

directly onto a keyboard. The letter usually needs to be drafted and then reviewed before the final version is printed, signed and released. This is where IT can be a particular help, as discussed in Chapter 7. The format of a letter is often determined by the house style of the company and the type of stationery used. However, there should always be an originator's reference number and a date. When you are replying to someone else's letter it helps to quote their reference and date.

Whenever possible address the letter to a particular named individual. The salutation ('Dear Fred') can then be made more personal, perhaps handwritten like the subscription ('Yours sincerely'). In the age of the laser printer a neat hand-written salutation and subscription can give some distinction to your letters. If you cannot write to a named individual, try to put a specific appointment such as marketing director. You may be writing to someone in your own profession, on a professional matter rather than a commercial one. For example, this might be to advise them of a professional meeting and invite them to attend. You can then address them as 'Dear colleague'. There are many detailed conventions which you get used to in particular jobs and situations. This book is not intended to give detailed advice about the layout and conventions of letter writing; you will find plenty of advice in the highstreet bookstores if you need it.

The technical report: This is a detailed and often lengthy report on a particular topic. The distinguishing feature is that the writer and all likely readers have a shared context which includes this particular technical discipline. You are writing for specialists, or at least for people with a general awareness of the field, so that they can fairly easily be lead into the detail of what you have to say. Apart from this one feature, a technical report is similar to a management report (see later), and needs to be structured and presented accordingly. A technical report is likely to include more data, analysis and perhaps pictorial presentation than most management reports. The IEE publishes a useful pamphlet on technical report writing; see IEE (1993).

The management report: This is the most complex type of written business communication. It is written for nonspecialists, often general managers (recall the definition in Chapter 1). It is on a complex topic. It may include data, analysis and recommendation.

The writing of management reports is considered in detail in the next two sections. Not only are they the most important and challenging type of written business communication, but they incorporate general principles which should be used for all the other, less formal, types of written communication.

First, though, there are some topics which apply to all types of written communication. The first is the use of figures and diagrams. These can be very helpful, but it is worth considering exactly why 'a picture is worth a thousand words'. Look at Figure 4.1. They are all familiar items. The objects are from the office, and the map is a simple one of Scandinavia. But just think for a moment how you would describe those items, in words, to someone who was not familiar with them. You could take several hundred words describing even the paperclip

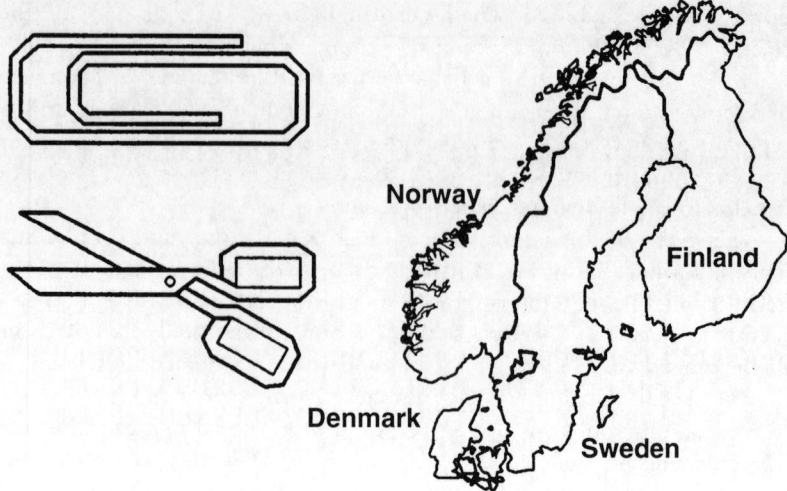

Figure 4.1   Diagrams

unambiguously. To describe the coastline of the map precisely would be impossible. The problem, of course, is that written language is just a special form of spoken language. It is essentially serial, or one-dimensional. It is good for telling a narrative sequence in chronological order. But it is not good at describing relationships which are not in that order. Thus it is not good at representing two- or three-dimensional objects unless they are familiar or are built from a few familiar components. A map is the supreme example of the efficiency of a diagram. Think how much clearer it is to give a two-dimensional (x-y) grid reference for a destination, rather than attempt to describe how to reach it through a maze of roads from a known starting point.

Engineers will need no further reminding of the value of figures and diagrams to support written or spoken text. They will also be familiar with the analysis and presentation of quantitative (numerical) data. Often, though, it is the final stage of analysis and presentation which requires the most careful thought. It is illustrated in Figure 4.2. Consider an example of this. Table 4.1 (on page 62) is a simple dataset recorded in a spreadsheet on a PC. It shows the electricity consumption for a particular building each month over a period of three years. The primary data are in columns C and D. These are the meter readings for day (full rate) and night (low rate) tariffs. Column E then calculates the day consumption for the month, column F the night consumption, and column G the total. The consumption is measured in kilowatt-hours.

As an engineer you will be used to scanning data sets like that, to look for significant patterns. In this case, of course, you would expect a seasonal variation through the year, with smaller variations between the same month in successive years. That much is evident from a quick perusal of the data.

But how would you present this information to a manager or an accountant in your business? The first rule in planning a business communication is to ask

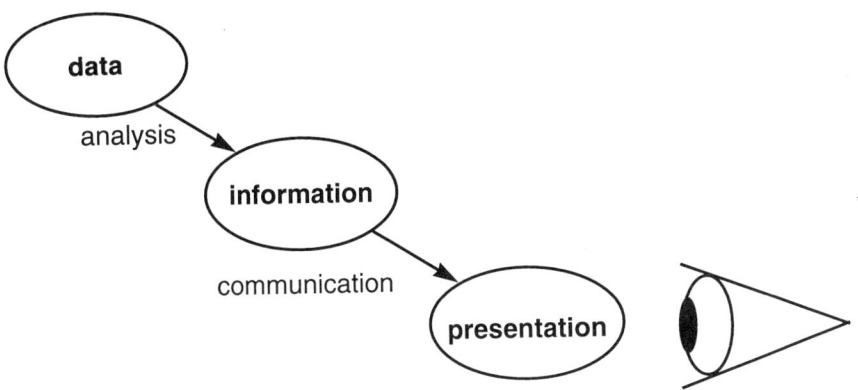

Figure 4.2   Data analysis

yourself what the recipients want or need to know. They are not interested in kilowatt-hours; they are interested in the financial cost. So, get the spreadsheet to calculate costs by feeding in the day- and night-tariff rates and calculate the total cost of electricity, month by month. Then present it as a bar chart with the three years arranged so that you can compare the cost for the same month in successive years. This is shown in Figure 4.3. This is a more useful method of presentation; it has turned the data into management information relevant to the interests of the recipients. You would supplement the bar chart with the total costs

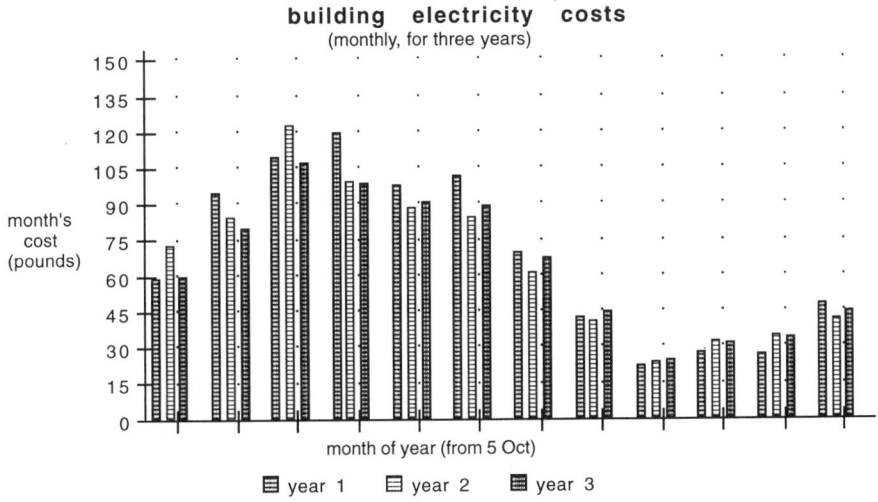

Figure 4.3   Three-year histogram

Table 4.1   Data set

Row \Col:	B	C	D	E	F	G
2	TABLE 4-1:					
3	Electricity Consumption Data-set					
4	(All figures are in kilowatt-hours kWh)					
5		-READINGS-		-----UNITS-----		
6	DATE	DAY	NIGHT	DAY	NIGHT	TOTAL
7	--					
8	05-Oct-92	6074	27908	0	0	0
9	05-Nov-92	6547	28781	473	873	1345
10	05-Dec-92	7339	30155	792	1374	2167
11	05-Jan-93	8308	31669	969	1514	2483
12	05-Feb-93	9318	33276	1010	1607	2617
13	05-Mar-93	10093	34667	775	1391	2167
14	05-Apr-93	10950	36230	857	1562	2419
15	05-May-93	11463	37506	513	1277	1789
16	05-Jun-93	11797	38236	334	730	1064
17	05-Jul-93	12016	38558	220	322	542
18	05-Aug-93	12291	38886	274	328	602
19	05-Sep-93	12543	39205	252	319	571
20	05-Oct-93	12958	39877	415	672	1087
21						
22	05-Nov-93	13557	40986	599	1109	1708
23	05-Dec-93	14280	42243	723	1257	1980
24	05-Jan-94	15400	43784	1120	1541	2661
25	05-Feb-94	16243	45228	843	1444	2287
26	05-Mar-94	16952	46517	709	1289	1998
27	05-Apr-94	17598	47917	646	1400	2046
28	05-May-94	18089	48842	491	925	1416
29	05-Jun-94	18407	49449	318	607	925
30	05-Jul-94	18654	49742	247	293	540
31	05-Aug-94	18990	50081	336	339	675
32	05-Sep-94	19342	50423	352	342	694
33	05-Oct-94	19702	50993	360	570	930
34						
35	05-Nov-94	20171	51968	469	975	1444
36	05-Dec-94	20862	53113	691	1145	1836
37	05-Jan-95	21808	54499	946	1386	2332
38	05-Feb-95	22658	55845	850	1346	2196
39	05-Mar-95	23501	57086	843	1241	2084
40	05-Apr-95	24248	58414	747	1328	2075
41	05-May-95	24744	59577	496	1163	1659
42	05-Jun-95	25119	60251	375	674	1049
43	05-Jul-95	25329	60699	210	448	658
44	05-Aug-95	25649	61049	320	350	670
45	05-Sep-95	25979	61441	330	392	722
46	05-Oct-95	26343	62133	364	692	1056

for each year. At current prices, rounded to the nearest pound, these are £830, £797 and £790. Thus the average annual cost is £806. Always remember to say whether historical prices have been adjusted, to allow direct comparison in real terms. In this case they have.

The managers and accountants have to manage cash flow in the business. They might therefore be interested in the profile of expenditure on electricity through the financial year. In the UK this ends on 5th April each year. You therefore get the spreadsheet to calculate the average cost, over the three years, for the first quarter (starting 5th April), the second (starting 5th July), the third (starting 5th October) and the fourth (starting 5th January). The figures come to £204, £91, £198 and £313 respectively. Always do simple cross checks; in this case the sum of the four figures (£806) agrees with the average annual cost previously calculated.

How should you present those four figures, to make the clearest comparison? There are several possibilities:

1 Present the four costs, and their total, as part of the narrative text in your report.
2 Prepare a pie chart which shows the relative size of the four costs. Your spreadsheet will probably produce something like Figure 4.4.

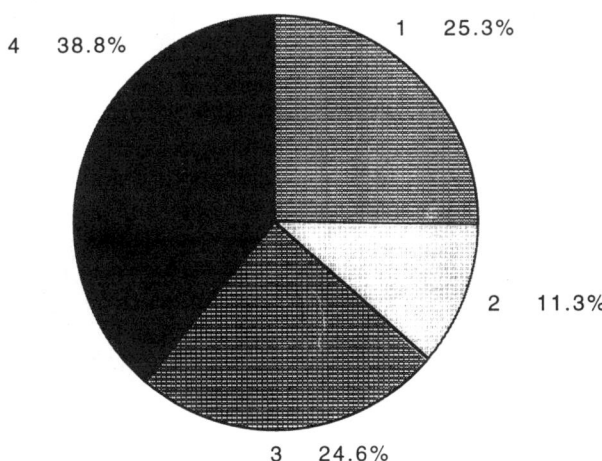

Figure 4.4   Data presentation: pie chart

3 Prepare a bar chart which shows the absolute size of the four costs. Again, your spreadsheet will produce something like Figure 4.5.
4 Prepare a graphic/diagram which shows the relative size of the costs, in proportion to the area of four circles. This is shown in Figure 4.6.

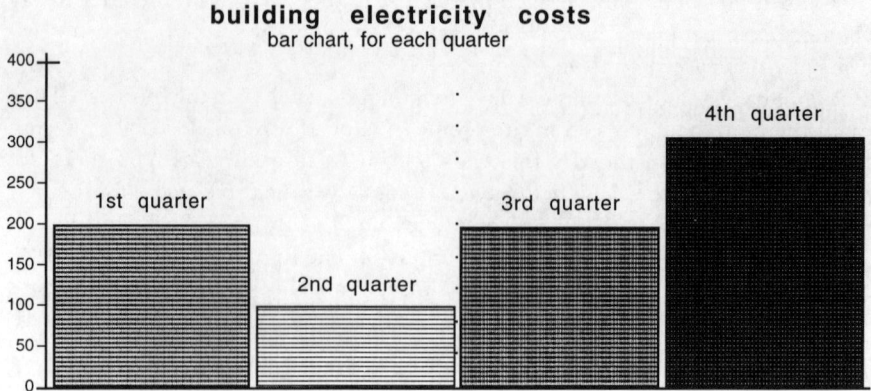

Figure 4.5   Data presentation: bar chart

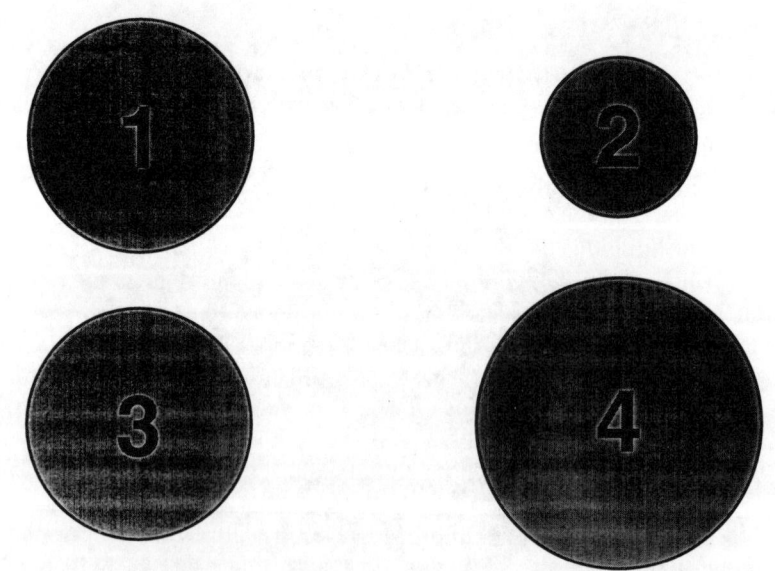

Figure 4.6   Data presentation: circles

5   Prepare a graphic/diagram which shows the relative size of the costs, in proportion to the area of four squares. This is shown in Figure 4.7.

6   Prepare a graphic/diagram which shows the relative size of the costs, in proportion to the volume of four cubes. This is shown in Figure 4.8.

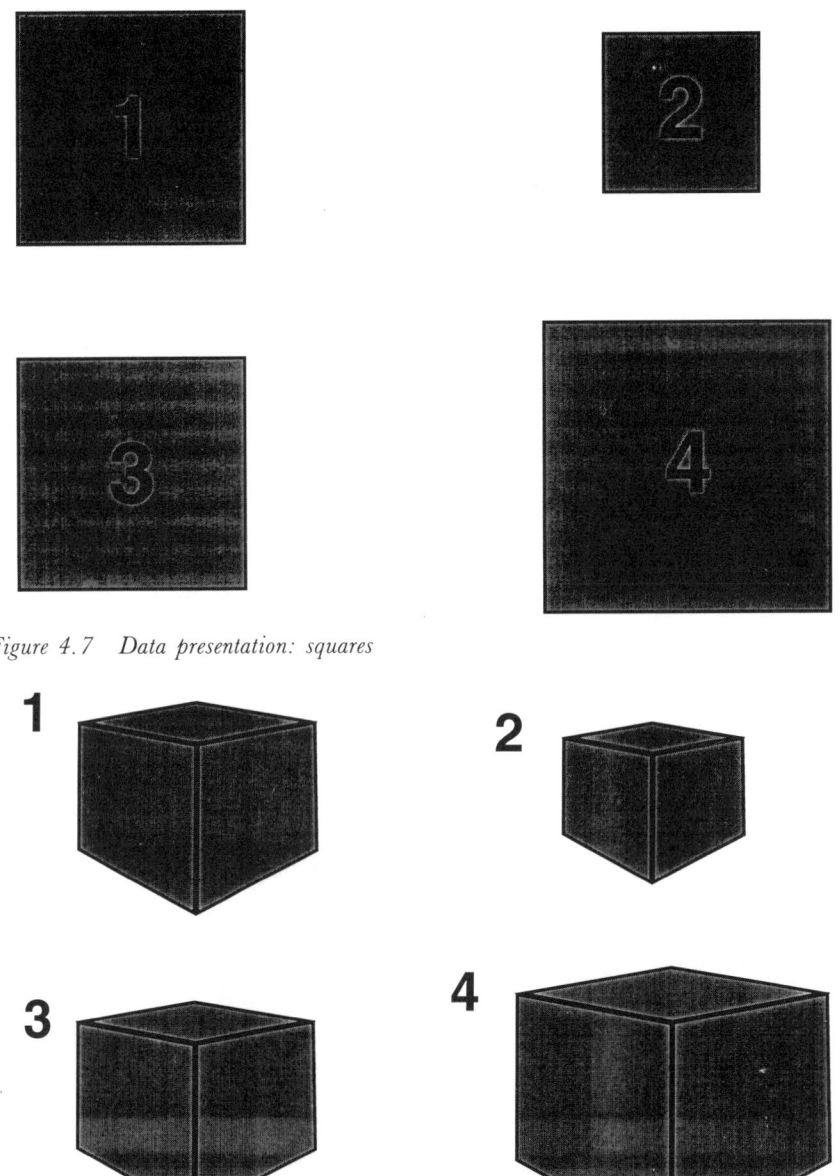

Figure 4.7 Data presentation: squares

Figure 4.8 Data presentation: cubes

Which method of presentation is best, and why? Think about it. In a simple case like this, listing the four costs in the narrative report would probably be adequate. But if you wished to portray them graphically there are several general points to note:

- Only one graphical method (no. 3) gives the absolute costs, and even that would not be precise enough for financial planning. In all cases, therefore, you would need to give the actual figures.
- Method 3 (the bar chart) does give a clear impression of differences, in absolute terms.
- Method 2 (the pie chart) is difficult to assess. Without the percentage figures (inserted by the spreadsheet) it is very difficult to assess the relative sizes by eye. Pie charts are not very good.
- Methods 4, 5 and 6 rely on the eye judging the relative size of standard geometric shapes. Modern graphics packages tend to use these, in a fairly decorative manner, but what does the viewers eye/brain say? It tends to judge linear distance better than area. Thus methods 4 (circles) and 5 (squares) make you underestimate the differences.
- This effect is even more marked for judging volumes, as in method 6 (cubes). Modern 3-D graphics packages tend to use such effects a lot. They may be attractive, but be aware that they can be deceptive!

This example is a simple one, but the principles of fair and accurate presentation of quantitative data to nonexperts are important. You must be fair, objective and helpful to the reader.

4.3 Report writing: the principles

If you know how to write a good management report you are well placed to structure and deliver most other types of business communication as well. The same principles apply.

Section 2.2 looked at examples of spoken and written communication. In both cases there was a one-to-many mode and a one-to-one/few mode. A fictional story in a magazine is an example of the first; a management report is an example of the last. To sharpen this distinction, consider what each of these examples of written communication is trying to achieve.

It is closely related to the way we think. It is well established that different parts of the brain are concerned with different types of mental process. One of the basic distinctions is between what is done in the left hemisphere of the brain and what is done in the right hemisphere of the brain. Buzan (1988) lists the left-brain activities as logic, lists, linearity, words, numbers, sequence, analysis and other similar activities. The right-brain activities are rhythm, colour, imagination, daydreaming, dimension, spatial awareness, music and other similar activities.

In summary, the left brain controls the right side of the body, and activities like speaking which require precise movement of the muscles. It also deals with rational thinking in a sequential manner. Here are some examples of left-brain thinking:

- doing a mathematical calculation
- controlling complicated machinery, such as a car

- solving problems using a systematic procedure
- planning a project, and reviewing progress against targets

In contrast, the right brain controls the left side of the body, and intuitive thinking in a parallel (holistic) manner. Here are some examples of right-brain thinking:

- wondering how much you like somebody
- appreciating spatial relationships, as in architecture and art
- solving problems by using intuitive judgement
- trying to persuade someone about your values and beliefs

We all do both left- and right-brain thinking. There is a physiological basis to the distinction, although the two hemispheres of the brain are connected by a data highway which co-ordinates their work. Recently it has been found that some tasks, such as complex linguistic problems, are processed differently by men and women. Men tend to use only their left brain for these particular tasks, whereas women use both left and right brain. This accords with the common observation that men tend to be more logical (left brain) and women more intuitive (right brain) in their approach to certain problems. It also relates to the distinction between task and process drawn in Section 3.2. The task ('the words') requires left brain thinking; the process ('the music') requires right brain thinking. This may be one reason why engineers (who of necessity are strong on left-brain thinking) often find dealing with the softer aspects of process difficult. The increasing number of women engineers may be better placed for that. Their skills can often bring a welcome balance into a task-oriented and male-dominated situation.

Figure 4.9 summarises the distinction between left- and right-brain thinking, and relates it to two types of writing. Take the fictional story (one-to-many) first. It is written for our enjoyment. It will involve our emotions, and situations which

LEFT **BRAIN**	facts logic sequential detailed	feelings emotions parallel holistic	**RIGHT** **BRAIN**
non-fiction (report): **data** **+ analysis** **= information** **conveys knowledge** **for decision / action**			**fiction (story):** **character** **+ situation** **= action** **conveys emotion** **to give pleasure**

Figure 4.9   Thinking and writing

contrast with our everyday task-oriented duties. The emphasis is on people (characters) who are portrayed in particular situations. Those situations are often chosen to contrast vividly with our usual experience: they are dramatic, challenging and may be far removed from our normal life (as for example in science fiction). The response of the characters to the challenge of the situation is shown in their actions. These arouse in us feelings of sympathy, fear or admiration. This is what we enjoy from the fictional story. It is also what we enjoy from an exciting, true-life story which is remote from our own experience. In either case the story conveys emotion to give pleasure. Reading it for pleasure is a right-brain activity. Analysing it for a literature exam would be left brain, however, and for most people not as enjoyable!

The management report is quite different. It is one-to-one/few. It starts with facts (data), analyses the data to create management information which will be used, along with other knowledge, as a basis for decision and action. Remember the management learning cycle which was discussed in Section 2.4, with its underlying value chain of information.

This distinction between left-brain and right-brain thinking, and the corresponding types of written material, is worth bearing in mind. As you read material in your business and private life, consider which category it is. What is the balance between logic and emotion? In business, it will mostly be about logic; in newspapers you will get both; in leisure it may be mostly about emotion.

The first principle of writing a management report is to follow the motto established in Chapter 3 for the use of language: be logical where you can; be persuasive where you can't. As with spoken communication, this does not mean that reports should always be merely cold and logical. Sometimes it is appropriate to convey commitment and enthusiasm. But these more emotional aspects should not be the dominant feature. Moreover, they should not be tangled up with the logical part of the report. Often they appear briefly at the end.

The second principle derives from the latter part of Section 2.4. You need to establish the purpose, aim and scope of the report. For any substantial business communication, like a report or presentation, you need to decide (and state):

- *Purpose:* the reason for making the communication. This usually relates to solving a management problem.
- *Aim:* what the communication itself tries to do. This is usually to contribute in a specific way to solving the problem.
- *Scope:* how much of the information value chain the communication deals with.

The purpose defines the context; the aim defines the contribution within that context; the scope defines how far along the information value chain you intend to go. Some practical examples were considered, relating to management reports. At this point you may wish to go back to Section 2.4 to remind yourself.

Remember, the reader is your customer. You can only establish the purpose, aim and scope of the report in consultation with the intended reader. You should actively try to understand the customer's point of view. Like all customers, they may not know what they want, or need. It is your business as the supplier (writer of the report) to help them define that. You must take a proactive role. There are several situations which can arise:

— The principal reader (perhaps your boss) has asked you to write a report on a particular topic. The instructions may have been verbal or in writing. In either case, try to write down the purpose, aim and scope of the report which you believe is required. Then consult with the reader, to see that you have understood the requirement properly. You may feel that a wider scope is needed; you may feel the report would not address the real cause of a problem; you may need extra resources to research and write the report. Whenever possible, reach agreement with the reader about the purpose, aim and scope. It is a good idea to confirm that in writing before you embark on the detailed work. It will help clarify your mind, and help avoid any misunderstanding with your customer.

— The topic and report are particularly complex. In this case you must reach agreement about terms of reference (TOR) for the report. Make sure that every word is right, and contributes to the intended meaning. This is because the TOR will appear in your report (perhaps as an Appendix) to clarify or define its aim.

— You are writing the report on your own initiative. This may be to highlight a problem and hopefully contribute towards its solution. Even in this case you should try to raise the matter beforehand and agree the purpose, aim and scope with the person to whom you have in mind to send the report. It may be that they know more about the situation than you thought. It is no good sending an unexpected report which deals with the wrong problem.

So you must get the purpose, aim and scope clear before you go any further. Write them down, if only for your own benefit at this stage. You should then think about the reader, and the management situation, from a slightly different angle:

● What is the reader's mind-set? Consider whether they may have preconceived ideas or prejudices; whether they have any personal preferences about the style or format of a management report; whether they have any preference about how you set about the task and present the results. In doing this you are using your own judgement about the reader, to supplement what you have formally discussed with them.

● What is the shared context which you have with your reader? Relate this to the model of communication which we discussed in Section 2.2 (Figure 2.4). Are you both from the same professional background, so that you are both familiar with the general terminology of that discipline? How much does the reader already know about the problem or issue which you will examine in the report? This could affect how much background information you need to give in the introduction, and the level of knowledge which you can assume throughout the report.

● What is your reader's purpose in asking for the report and then using it? Try to think yourself into their position and job. What do they need to find in the report? What do they want to find? These two questions may have different answers! What will the reader do as a result of reading your report? Is that actually in the best interests of the business? How should this affect the way you write the report?

The purpose of these questions is to think about the shared context and purpose of the report from a wider perspective. Even if you have already formally agreed the purpose, aim and scope with the reader this may not be the whole story. There are issues of personal and organisational politics which may be relevant. There may be other types of personal interest involved. There may well be several people who will read the report; think about them all, but especially the person who is in a position to take the executive decision which needs to follow the report. You would be naive to ignore these factors, even if you cannot state them openly. They can and should influence the way you approach the task and present the result. They may even cause you to amend the words of the purpose, aim and scope which you have prepared as your own working document.

That second principle of writing a management report derived from the overarching rule: the reader is your customer. The third principle also derives from the same rule. It is that your report should have multiple levels of detail. This means that a reader can get your message easily and quickly at the broadest level, or can choose to read your report at several other levels, progressively more detailed. Here are some reasons why this is helpful to the customers:

— If your report is on an important topic it is likely to be read by several people. Some will only have a specialist interest; they may be checking that what you are saying does not conflict with established policy in their specialist areas. Some will be senior people, who will only have a short amount of time to spare; it is important that they can get your message easily and quickly, especially if they are called upon to make a decision based on their reading of it.

— Some people will need to read the report in close detail. The person who called for the report (perhaps your boss) is probably one of them. However, even these people may not want to get all the detail the first time they pick the report up. Because of the principles of human information processing discussed in Section 2.2, it is always a good idea to have several bites of the cherry: having a quick skim of the report, then reading it more closely on subsequent occasions. If your report has multiple levels of detail, this process will be easier and more effective.

— Your report may well need to be kept as a record of the analysis or rationale which underpinned a management action. Therefore you as the writer, or any of the readers, may need to come back to the report at a later date. This may be as a reminder of the contents, or to check whether detailed facts still apply.

For all these reasons, a management report that is clearly structured to offer multiple levels of detail will help its readers. Furthermore, it will help you to plan the writing of the report.

The fourth and last principle is that your report should separate fact, analysis and opinion. That sounds fairly obvious and desirable, but it is amazing how often people mix these three things together. It is then very difficult to sort them out. In particular, if the reader does not agree with the analysis or opinion then it is very difficult to amend the report to build effectively on what is already there. There are two main reasons for separating fact, analysis and opinion. The first is to help you to plan and present the report logically. The second is to help those readers who want to develop their own analysis and interpretation of the facts which you present. They may even give you some guidance and ask you to do it for them! In that case you will be saving yourself a lot of extra work when you come to amend your report.

In summary, then, the principles of management report writing are:

1   Be logical where you can; be persuasive where you can't. This governs how you use language and structure your report.
2   Agree with the customer the purpose, aim and scope of the report. This means doing the groundwork with the customer before you start writing.
3   Offer multiple levels of detail in the report. This will help the greatest range of readers, on a range of occasions.
4   Separate fact, analysis and opinion in the report. This helps you, as well as them.

The first principle relates mainly to the way you will use language in your report. The discussion in Chapter 3 is relevant here, with an even greater emphasis on logic in this written case. The second principle has been discussed already in this section, notably the way you need to negotiate with the customer. The last two principles relate to the report itself. These are discussed in the following section.

4.4 Report writing: the practice

You have now done all the groundwork: defined and agreed the purpose, aim and scope of the report; and thought about the wider context and people involved. You now reach that agonising moment when you know you have to start writing the report. The blank sheet of paper stares reproachfully at you. The PC screen is empty apart from the remorselessly blinking cursor. What do you do next?

Well, you mustn't just start to write the report! You must plan its structure in some detail first, otherwise you are liable to waste a lot of time. How can one devise a structure, and write within it, so as best to satisfy the last two principles defined in Section 4.3? These were, to offer multiple levels of detail, and to separate fact, analysis and opinion.

To help, it's worth thinking about how similar requirements are met in the quality newspapers and periodicals that we read:

—   Fact, analysis and opinion are presented in a different style, and usually on different pages of the newspaper. For example the factual news items may be in the first part of the paper. Analysis and comment is often in regular columns by named individuals; you know that you are going to read their interpretation and comment on the facts of the recent news. Editorial comment from the newspaper itself is prominently marked as such, usually by placing it as a leader on pages near the centre. Thus each of the three categories is clearly flagged, and you can choose what you want on different occasions. Often people want straight topical facts (news items) first, and then will spend more time reading and thinking about the comment and opinion. Some articles, often in the supplements, are features: they may include both fact and opinion. But if you examine the item in detail you should be able to distinguish between them.

— Most items offer multiple levels of detail. This is necessary because not all readers will have the same level of interest in each item. Some people will skip whole pages of a newspaper, knowing it is unlikely that there will be anything there to interest them. This might apply to the sports pages, the business pages, or even the foreign news. On other pages they might just skim the headlines. The headline can have two purposes:

- For a news item it should give the basic facts as briefly as possible. This might be designed to attract the reader to find our more ('Spy scandal rocks government') or be boringly factual (like the legendary 'Small earthquake in Chile: not many hurt'). Below the headline the news item is structured in so called pyramid writing. This means that successive sentences and paragraphs contain progressively more detail on the same topic. You simply read as far as you wish to and then stop. The key points are at the beginning rather than at the end.

- For a comment or feature item it should be as attractive as possible. It may even be enigmatic, so you are not sure what the item is about until you begin to read it. It arouses your curiosity and interest. When you look in detail at the item it usually has three main parts: an introduction which leads you into a statement of the topic to be discussed; the body of the article which discusses various facets of the topic; and a conclusion which summarises the main points or the viewpoint which has been argued.

Look at the newpaper or journals you read to see which technique they use and where. Consider how many of them apply to our problem of achieving the two similar criteria for a management report.

The closest analogy for the management report is with the feature article. The three-part structure (introduction, body and conclusion) is the basis for most written business communication. It has been summed up as: say what you're going to say; say it; then say what you've said. This leads readers into what we have to say, and achieves the necessary reinforcement of key points, as discussed in Section 2.2.

Figure 4.10 suggests a more complex structure which will meet the more demanding requirements of a management report. Please note that the structure proposed is provisional. It is a template or starting point which you should adapt to meet the needs of each particular situation. However, it should help you meet

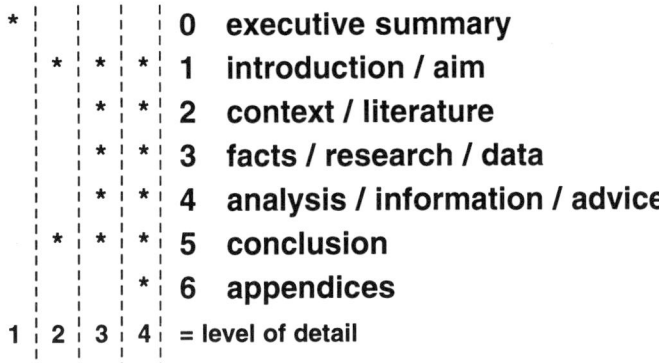

Figure 4.10 Provisional report structure

the key requirements: to offer multiple levels of detail, and to separate fact, analysis and opinion.

The diagram shows seven parts to the report. There will also be a title, of course. Three of the parts (2–4) make up the body of the report. By reading the appropriate parts of the report, the reader can choose between four different levels of detail. These are shown by the asterisks in the four columns. The four levels of detail are:

1 Executive summary (or abstract)
2 Introduction + conclusion
3 Introduction + body + conclusion
4 Introduction + body + conclusion + appendices

The executive summary will be placed conspicuously at the front. Everyone should read it, even if they intend to go into greater detail afterwards. Experienced readers will have developed their own technique for doing that. Most will skim through the contents and the document itself to get a feel for its structure. Everybody who moves beyond the executive summary should read the introduction. If you feel there could be doubt about the options, you could explain them briefly there, especially the option of reading the introduction and the conclusion as a coherent summary of the whole report (level 2).

Look in more detail at each part of the report:

> *Title:* The title is the ultimate summary of your report! It should accord with the aim of the report (see later). Every word should contribute to the overall meaning. Although you are not writing a novel, try to make the title attractive. It should give an immediate impression of what the report is about, and give people a good clue as to whether they need to read the report at all. Sometimes it is not possible to do all this in a title of reasonable length. In that case consider a subtitle. Management books (like this one) often do that.

Executive summary or abstract: Always include this, whether you are asked for it or not. It is helpful for you and for the readers. It makes you distil the essence of the rest of the report. It can help to expose weaknesses in your argument, or in the way you present the material. For the readers it is the important first encounter with your report. Some readers will go no further, so it is important that you get the message across to them clearly. They are likely to include the more senior people who may have to take decisions based upon your report. It is vitally important to get it right!

Introduction/aim: The first chapter (or main section) of your report can usually be called the introduction. It gives the background and sets the scene as to why the report is being written: the purpose which you have already agreed with the main reader. To decide how much to include, you need to consider the shared context you have with the likely readers: how much they are familiar with the situation, and how much general awareness they have of the topic. If in doubt, include more background than you think is strictly necessary. Even if it is familiar, it will help focus the reader's mind on the weightier material to come. This part of the introduction should lead naturally to the statement of the aim, which you have also agreed beforehand. It should be stated explicitly: 'The aim of this report is to...'. This should be the most carefully crafted sentence in the whole report. Finally in the introduction, go on to explain the structure of the rest of the report. This gives a mental map to the reader. It will accord with the scope which has been agreed beforehand. Mention the conclusion, and the fact that it will make sense, as a summary, if read straight after the introduction.

Body: The body of the report nearly always comprises several chapters (or main sections). Never use the term 'body' as the title, even in the exceptional case where it has only one chapter. The body contains the meat of your report. It will separate fact, analysis and opinion. This can be done by having separate chapters for each, or by separate sections within each chapter. In the example of Figure 4.10, Chapter 2 would have detailed background information about the problem situation and information or knowledge from relevant literature or other sources. Chapter 3 would assemble the detailed facts, and the data from any research investigation carried out. Chapter 4 would analyse the data into management information, reach conclusions and make recommendations (if this was within the agreed scope). In this case therefore the analysis and all decisions requiring judgement would be concentrated in Chapter 4. You should make specific reference to all supporting appendices (see later).

Conclusion: This should be a summary of the body of the report, containing no new information. It has two main roles:

- For a reader who has just read the introduction and the body, it serves as a recap of the detail just studied. It completes the level 3 reading of the report.
- For a reader at level 2 (who has come straight from the introduction), it completes the report in a coherent manner.

You need to have both these readers in mind when you write the conclusion. In a complex report, it can be useful to tabulate the main conclusions, with a cross reference to the place in the body of the report where the supporting detail can be found. This helps a reader to check the arguments or discussion which led to a particular conclusion. Similarly, recommendations for decision or action can be tabulated. Some people like to have recommendations in a separate, final chapter (or major section); that does not affect the principle.

Appendices: These give supporting detail to the body of the report. Detail is placed in an appendix when it is expected that few readers will need to study the detail closely, or it would disrupt the smooth flow of the main part of the report. Sometimes an appendix is like a minireport in its own right, leading to conclusions which justify assumptions stated in the body of the report. Sometimes an appendix is written in a pyramid style; the reader simply carries on until they have all the information they need from the appendix. Remember that the pyramid style is not right in the body of the report! There are some important points:

— Every appendix should be mentioned explicitly in the body of the report, with an indication of what it contains. You must say enough to enable the reader to decide whether they need to look at that particular appendix or not. The title of the appendix should be a further help to this.

— List the appendices with their titles at the end of the contents page of the report. They should be numbered (or lettered) in the order in which they are cited in the body of the report. A different order will tell the reader you have reshuffled them during your drafting of the report, and can't now be bothered to put them into a tidy order.

— In complex cases, appendices may need their own supporting documents. You might call these annexes. Number or letter them to avoid confusion with main appendices.

The provisional report structure shown in Figure 4.10 can therefore meet the key requirements, provided you follow the principles mentioned. To consolidate this, take as an example the third (and most detailed) of the reports whose purpose, aim and scope were considered in Section 2.4. The purpose of the report was to define the morale problem, and suggest how it might be solved. The aim was stated as 'The aim of this report is to identify why morale is low, and recommend how to improve it'. The scope of this report was the most comprehensive, in terms of the information value chain shown in Figure 2.9. It therefore had to

● provide data (facts) about the situation
● analyse that data to provide information relevant to the problem
● provide relevant knowledge from a wider, or related, context
● recommend a decision, based on that information and knowledge
● say how the related action can be initiated, and by whom

Think for a moment about how you would structure the report if you had to write it. Jot down your ideas, before you compare them with the following suggestions.

Title: 'Steps to improve morale at ABC Manufacturing Co Ltd'. This says which company one is considering. The phrase 'Steps to. . .' indicates that the report is about doing something. Note that if we just said 'Improving morale at ANC. . .' it would be ambiguous: it could be reporting the facts of an improving morale situation, rather than recommending how we should deal with a poor morale situation. Make the title as unambiguous as you can.

Executive summary: the concise (two-page?) summary for senior people. Remember, this is not just a lead-in to the report; it is a summary of the whole report and its conclusions.

Contents: listing all chapters (or major sections), appendices and figures.

Chapter 1: Introduction: Brief background to the company and its current situation; symptoms of poor morale; how they came to notice, and when; likely impact on people and their performance; why we need to do something about it. This leads into the explicit statement of the aim: 'The aim of this report is to identify why morale is low, and recommend how to improve it'. Describe briefly the structure of the rest of the report, and the option to go straight to the conclusion for a quick understanding of the report.

Chapter 2: Evidence of poor morale: This would give the factual basis of the situation. You could assemble data about absenteeism, productivity levels, participation in internal communication and suggestion schemes, and staff turnover. Some of these data would probably go into supporting appendices, with the key facts included in this chapter. You could include information from standard texts on human resource management (HRM) or from HRM specialists in the firm.

Chapter 3: Employee attitude survey: You might decide that the best way to get to the root of the problem would be to conduct an attitude survey. You would involve the HRM specialists in this. It would yield data about how people felt about the firm and their working conditions, and what they felt needed doing about them. Chapter 3 would describe the methodology for this research, when it was conducted, and the results. A copy of any questionnaire, or the format for structured interviews, would be attached as appendices. Detailed results would probably also be in appendices, with the key results in the main text.

Chapter 4: Analysis and recommendations: This chapter would be more discursive. You would stand back from the detail, and relate it to the overall problem and aim as stated in Chapter 1. You would identify the key problem areas. Some might emerge clearly from your analysis of the data (e.g. everyone says they are underpaid); others may require your experience to discern (e.g. the root cause is not pay but the boring nature of most of the jobs). From this analysis you would formulate your recommendations. Remember to make them specific: say who should do what, and when. Avoid the passive tense; it obscures who should do things, and makes it more difficult to allocate responsiblities.

Chapter 5: Conclusion: This would summarise the whole body of the report (Chapters 2 to 4 inclusive). It would not normally refer directly to appendices;

if people want that detail they should reach it by going back to the appropriate detailed chapter first. It therefore helps to say what you dealt with in which chapter. Chapter 5 would end with a distillation of the conclusions and recommendations. This is often best done as a bullet-point list. You may decide to include cross references to the relevant earlier detail.

Appendices: Being diligent, you will include these in the order cited. Each should have a descriptive title. Make sure each is mentioned explicitly in the main text, and there is clear link. An example would be: 'Appendix 4 gives the detailed results of the attitude survey. They key complaints from the staff were about pay, quality control procedures, and constant interruptions. More than half the staff mentioned all three of these issues'.

Well, how did you do? Did your ideas for the report structure bear any resemblance to that? It would be surprising if it accorded in every respect, for we all have our own ideas for structuring and presenting information on a particular topic. But your structure should meet those key requirements: offering multiple levels of detail; and separating fact, analysis and opinion. The findings should be crystal clear at the end.

Figure 4.11 shows the 'shape' of a report which follows this suggested structure.

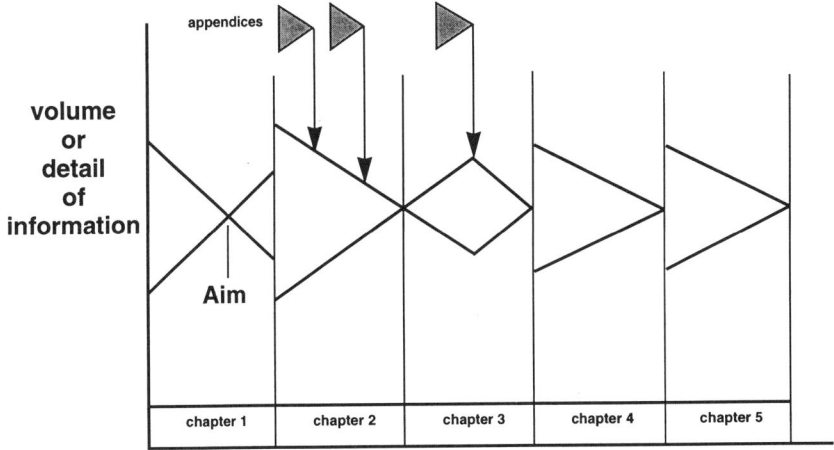

Figure 4.11   Shape of management report

The diagram echoes the discussion of the information value chain in Section 2.4 and Figure 2.9. The *y*-axis of the diagram indicates the volume, or detail, of information. See if you can relate the shape to the description of each part of the report above. It is subjective, but there are some general points to draw out:

● The introduction should converge from general background information to the explicit statement of the aim. It should then diverge a little, to describe the structure of the rest of the report.

- Some of the other chapters, and the appendices, will converge from a lot of data down to some specific information. Chapter 2 is an example of this case. Others will diverge then converge. Chapter 3, about the attitude survey, is an example.
- The conclusion should present an overview of the body and then converge down to the particular findings and recommendations.

The diverge–converge model can be a useful tool to plan your work; it complements the contents sheet which lists the chapters. It should help you adopt a logical approach to the problem and structure your report accordingly. You may find Figure 4.12 useful for that. It shows a general approach to problem solving, which you will find discussed in detail in many management books. It helps you to structure a report according to the principles that have been discussed.

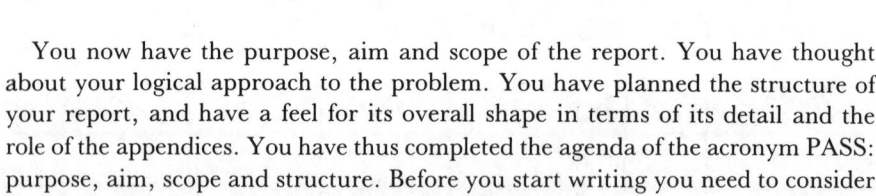

    1   **analyse the problem**

    2   **list the definitions and assumptions**

    3   **determine the criteria for the solution**

    4   **evaluate the possible solutions**

    5   **select the best solution**

Figure 4.12   Problem solving

You now have the purpose, aim and scope of the report. You have thought about your logical approach to the problem. You have planned the structure of your report, and have a feel for its overall shape in terms of its detail and the role of the appendices. You have thus completed the agenda of the acronym PASS: purpose, aim, scope and structure. Before you start writing you need to consider a few practical points about format and presentation:

> *Headings:* These help readers to follow where they are in the report, referring back to the contents page if necessary. They serve as milestones when reading the report, and help them to find specific topics quickly. Use a structured system of headings. The main title should be the most prominent; it is usually centred and highlighted with large font or a display box. Chapter or main section headings should be the next most prominent. Lower levels of heading should have their own particular style. It is a good idea to use a decimal numbering system. This book uses a two-level system; reports should use no more than three, usually. Decimal numbering is much more flexible when you are making changes during the drafting stage. So, draw up a decimal-numbered contents list before you start.
>
> *Page numbers:* In the final report, the page numbers should be added to the contents list. You can't usually do that until the final version is into the word

processor and has been paginated. So in the meantime make your cross references by decimal-numbered section. Sometimes the page numbers for appendices are numbered A-2 (page 2 of Appendix A), etc. Sometimes the whole report is page-numbered sequentially. The main principle is to ensure that no two pages ever have the same number! Most reports are in A4-portrait format; paper sizes for text and diagrams are discussed in more detail in Section 5.3.

House style: Your firm or organisation may have its own rules and conventions about layout and related matters. Learn and use them. Subject to that, always give figures a number and title, refer to them from the text, and list them in the contents (like appendices).

By now you are itching to pick up your pen, or keyboard, and start writing. Why have we spent so much time on the preliminaries? There are two reasons. One is to make sure that the report will serve its customer (the reader) well and make the intended contribution to management progress. The second is to save you time in writing the report. Writing a report is like planning and executing a project. In fact it is a project! Time spent planning is repaid fivefold by avoiding trouble later. So make sure you do all the preliminary spadework and planning suggested here. Then you will be mentally and physically prepared to write quickly and effectively. How do you approach that part of the task? That is the final topic in this section.

You don't start at the beginning (the executive summary) and continue writing until you reach the last element shown in your planned structure. Few people can leap from general to particular in the agile manner which that would require. Usually we are by now rather immersed in the detail of the work; it is difficult for us to stand back and present the overview needed, for example, in the executive summary. So we have to approach things in a different order. Here is one order which has been proved in practice by a large number of people:

Step 1: Make sure that you have indeed done all the preparation mentioned. You know the purpose, aim, scope and structure (PASS). You have the title and decimal-numbered draft contents list. You have a feel for where the detail will be, and especially the role of appendices. You are aware of the rules of house style to which you must conform.

Step 2: Write down the aim of the report. Check that every word contributes to its overall meaning. Make sure that the meaning accords with what the reader expects and needs. Compose a title which accords with the aim. If necessary use a subtitle.

Step 3: Draft the introduction. Recall Figure 4.11. You are providing background information, and leading readers into the formal statement of the aim. Then you are giving them a map of the rest of the report. Drafting the introduction is difficult; you are too immersed in the detail, and it is not easy to get into the swing of writing continuous prose. Take your time, and accept that you won't get it right first time.

Step 4: Draft, in order, the chapters or major sections which comprise the body of the report. Where the detail is to be supported by an appendix put the necessary linkage in the main text. Tell the reader what the appendix contains and how it supports your main argument. You may like to draft the appendices as you come to them, or all together at the beginning or end of this step. That is really a matter of individual choice, and can vary for each report. But make sure it is done before you move to the next step.

Step 5: Having done all the detail, go back to the introduction. It may now seem stilted or incomplete. Try to improve it, as a lead-in to the body of the report. Then draft the conclusion, making sure that it meets its two requirements: to summarise the body; and to round off the report for those who are reading only the introduction and the conclusion (level 2). That is the main reason for reviewing the introduction before drafting the conclusion.

Step 6: Now draft the executive summary. This is the most important part of the report. With the level 2 summary in your mind from the last step, you are now much better placed to draft this overall summary. Keep it to two pages or less. Make sure every word says something useful.

Step 7: You now have a complete draft of the report. Set it aside for a few days. This will only be possible if you have planned your work sensibly! Consider asking someone else to read the report and give you comments. It is best if they are not too familiar with the detail; you want to see whether the overview makes sense to such a person. An outsider (perhaps your partner) might be helpful for that, if the content is not confidential. If you are writing in other than your first language, ask a person whose first language is the same as that used in the report.

Step 8: Revise the draft, taking account of any comments from friendly readers. As always, try to put yourself into the viewpoint of the main recipient of your report. Pay attention to spelling, grammar, logic and presentation. These things can undermine the credibility of your report. Have you been logical where you can, and persuasive where you can't?

Step 9: Produce the final version of the report. Make sure there are enough copies, for a named distribution list. Does the report look as good as it is? Is it as good as it looks? Finally, submit the report by the required date. Await the reactions or response; there is always feedback sooner or later! Meanwhile, though, breathe a sigh of relief at a job well done.

This procedure may seem laboured and mechanical. But it is not; it is designed to help you produce an effective report in an efficient manner. You might query the need to set the draft aside for a few days, when the temptation is to get on and finish it. But when you come afresh you always see things which you can improve. Do not tinker endlessly, but make the changes which will really help the reader. A wise preacher was once asked how he prepared his weekly sermon. He said it was no good working from Monday to Saturday to prepare Sunday's sermon. Instead, he had six sermons in preparation all the time. He worked two hours on a different sermon each day of the week. Any one sermon would be started on Monday of one week and then be continued on Tuesday of the next, Wednesday of the next, and so on until it was given its final polish on the Saturday before the Sunday when it was delivered. The six two-hour periods of work were best spread over six weeks rather than concentrated into one week. The preparation required no more total time, but it was more effective when spread over a longer elapsed time. The same principle applies to drafting a report. The preacher also said 'There's a world of difference between loving to preach and loving those you preach to'. Similarly, don't get so bound up with the writing of your report that you forget who it is being written for, and why.

However, like the provisional structure (Figure 4.10), this procedure is meant for you to adapt in the light of your own experience. As you do that, though, make sure you adhere to the important general principles.

There is a final topic which concerns some engineers: writing for publication. Here is a general structure for an academic paper which is to be submitted to a journal for publication:

> Title
> Authors and affiliations
> Abstract
> Contents
> Introduction
> Literature review
> Hypothesis and research methodology
> Experimental procedure
> Experimental results
> Analysis and discussion
> Conclusions and recommendations (including further research)
> Bibliography

Always study your target journal carefully. Look at the papers in it, and the way they are presented. Follow any notes for authors; they may be printed in the journal, or available on request. It may be best to write to the editor saying what you propose, rather than sending a complete manuscript without warning. If you agree a deadline for submitting a manuscript, stick to it. Until you are experienced, allow twice as much time as you think you will need. After you have submitted the manuscript, don't pester the editor. Respectable academic journals require all papers to be sent to independent reviewers. This takes time, but when you get their comments treat them seriously. Revise your paper to meet their requirements, but without saying anything which makes you uncomfortable to sign as your own work. Sometimes it helps to submit a manuscript on computer diskette; check the formats required.

The bibliography/references should follow the journal's house style precisely. If you have a choice, I suggest the author (year) system, as used in this book. It is more flexible than sequentially numbered references if you need to rearrange material. For more information about writing for publication, see IEE (1992).

4.5 Summary

Written material in business is very important; it records ideas and decisions on major issues. As with spoken communication, we can distinguish between internal and external communication. Written communication must be accurate; we take responsibility when we sign it.

Section 4.2 identified the characteristics of several types of written business communication: memo, e-mail, letter, technical report and management report. The last is the most complex and challenging. It looked briefly at the benefits of using figures and diagrams; they are invaluable for portraying two-dimensional relationships. It then looked at the analysis and presentation of quantitative data. The example showed how the eye can deceive in graphical presentation of even simple figures. Modern graphics packages produce pretty diagrams which can be more deceptive to the eye than a simple bar chart. Exercise caution. That said, graphical presentation tailored to the needs of recipients can enhance the quality of a report significantly.

Section 4.3 moved to the principles of report writing, starting by noting the differences between left-brain (logical) and right-brain (intuitive) thinking. This is reflected in two contrasting types of writing: the management report and the fictional story. Their approach is quite different, and you should develop the habit of considering which applies in the range of written material which you encounter. A report should be purposeful and logical. The first principle of writing a management report therefore accords with that for business speaking: be logical where you can; be persuasive where you can't.

The second principle is to agree the purpose, aim and scope of the report with the principal reader. This is a two-way process; you are working towards a common view of the problem and how your report will contribute to its solution. Write down the purpose, aim and scope and wherever possible send it to the principal reader as confirmation. Then think about the wider aspects of the shared context and mind-set they may have.

The third principle is a requirement for the report itself: to offer multiple levels of detail. This will make it more accessible to a wider range of readers. It will help those who need different levels of detail on different occasions. It will also help anyone (including yourself) who may need to come back and look at the report in the future.

The fourth and final principle is that the report should separate fact, analysis and opinion. This will help you to structure it logically and will help readers who wish to develop their own interpretation of the facts, perhaps differing from yours.

Section 4.4 considered how to meet the last two requirements: multiple levels of detail; and separation of fact, analysis and opinion. The structure of the quality newspapers and journals provides an instructive example. We can use those techniques in a disciplined way. A provisional structure was suggested for a management report to help you do that. It offers four levels of detail and separates fact, analysis and opinion. It comprises title, executive summary, introduction, several chapters making up the body of the report, conclusion and appendices. These elements are structured and linked to meet our key requirements. We considered an example, and the shape of the resulting report.

Next, Section 4.4 considered the actual drafting of the report. You start with the purpose, aim, scope and structure (PASS). The nine-step procedure should then help you to do the work methodically, and also produce coherent summaries at the first two levels of detail (the executive summary, and the introduction with the conclusion). It is that overview which most writers find difficult to produce, after they have been immersed in the detail. You need to develop your own procedure, and then apply it flexibly in each situation.

Finally, it looked briefly at the special requirements for writing for publication in refereed journals, and a simple form of referencing.

This has been a long but important chapter. Remember that we have been talking about writing effective management reports, in a business context. The principles will apply in other situations, such as writing books. Those situations will have their own particular emphasis, but what has been said here should equip you to widen the scope of your writing.

In business, as elsewhere, the written word has power. Learn to use that power effectively and responsibily.

Chapter 5
Presenting it

'The art of speaking is made up of five things:
invention, arrangement, style, memory and delivery'
Alcuin of York (c735-804)

5.1 Introduction

You recall that there are two ways to read this book. After studying the first
two chapters you can either dip into selected chapters or you can carry on
reading the chapters in sequence. If you have the time, the second method is
better because some topics are relevant to more than one chapter. If you are doing
that you will now have completed four of the eight chapters in the book. You
recall that:

● Chapter 2 set the foundation for the whole book, with the key ideas about
  communication, the business enterprise and management.
● Chapters 3 and 4 covered important matters that can be applied to almost
  all types of business communication: the logical use of language; analysing
  and presenting quantitiative data; and the principles of structuring an effective
  business communication.

The three chapters that follow deal with special cases and situations:
presentations, meetings, and the use of information technology. This chapter
considers business presentations: their purpose, how to plan them, and how to
deliver them.

5.2 The purpose of presentations

Why do we have business presentations? Think for a moment about presentations which you have attended. Did you enjoy them, or did you endure them because you had to be there? How many were boring, and how many were engaging? Did you end up thinking 'So what?', or thinking 'Right; we now know how to get on with things'? What do you recall most? More likely than not, you remembered something out of the ordinary: a presenter with a colourful character (or even colourful clothing), a vivid style of presentation, effective use of humour, unusual visual aids, or a memorable demonstration.

This suggests that the visual channel is at least as important as the verbal channel (discussed in Section 2.2). You can probably recall plenty of images, but how many sentences can you remember? There are very few lectures or presentations from which I can recall key phrases verbatim. The notable exception was a supremely boring lecture which ended with the key phrase being repeated six times before we were released from our agony. The presenter achieved something, but his technique was appalling.

As with written communication, it is useful to distinguish between internal and external presentations. An internal presentation is made to one or more colleagues in the same part of the business enterprise. An external presentation is one made to outsiders (notably customers), whether they be people in a different part of the firm (with whom you may have a supplier–customer relationship), or people completely outside it.

Here are some examples of internal presentations:

1   You have been studying the detail of a new technology which could be relevant to the work of your department. You assemble a group of colleagues to tell them about it and invite their comments.
2   You have just attended a management or technical course. You have been asked to tell a group of colleagues about it. The firm wants to share around the experience you have gained, and decide whether it should send more people on the course concerned. The training officer will therefore be attending.
3   You are a member of an R and D team. You have just completed some work, and need to tell the rest of the team about it. Together, you will then be able to decide how to use that work further.
4   Your production department needs to tell your marketing colleagues about some possibilities for improving current products. You want to know whether this would be worthwhile from the marketing viewpoint.
5   Your firm has to reduce costs, and your department has to take its share of the misery. You have to address a meeting of your staff, to tell them what is going on and the implications for them.

Here are some examples of external presentations:

6   Your department has recently submitted its five-year plan, with the detailed budget for the next financial year. You have to persuade senior management

(and the accountants) that you understand what you are about. They want you to become a profit centre, rather than a cost centre within the firm.

7   The marketing director of your firm wants to know why the rate of production of the firm's new product cannot be increased faster. He wants to meet the demands of a buoyant market. He keeps talking about 'competing on time', but you are not sure what this means. He has asked for a presentation so that he can understand your difficulties and explore ways of overcoming them.

8   Your firm is bidding for a contract to supply engineering goods or services. You have to contribute to the firm's presentation, to explain to the customer the design and rationale for the product.

9   One of the firm's products is proving unreliable in operation. You have to prepare a presentation to a dissatisfied customer, to explain what went wrong and what you intend to do about it. The sales staff are anxious about this and are breathing down your neck.

10   The industry in which your firm works is regulated by several statutory bodies. One of these is concerned with the control of pollution. They have asked about your plans for future levels of waste disposal from your production site. Your managing director has asked your section to prepare a presentation about this, to give to the officers of the regulatory body.

Think about these ten situations in relation to the model of human communication which we discussed in Section 2.2 and Figure 2.1. In every case (except possibly the first) you as the presenter will have a different mind-set from your audience. You therefore need to consider the shared context carefully, to make sure you pitch the presentation at the right level and use only language which they can understand. If necessary, define a few key terms before using them in the presentation. This is even more important than for a management report; in a presentation the audience cannot stop to check the meaning of a new term. They will either sit silent and lose the thread of what you are saying, or they will interrupt with a question and disrupt things for you and the others.

Some of the examples are mainly about presenting the audience with new information, as a basis for a discussion which will follow on immediately. Which ones are like that? The others have the broad aim of supporting a product which you or your department passes to other parts of the firm, or to people outside the firm. That product could be a good or a service, including a management report. Which are like that?

Thus there are two broad roles for a presentation:

● to convey information, as a prelude to discussion
● to support a product, by explaining or promoting it

In every case, to plan an effective presentation you need to apply the principles of Section 2.4: define the purpose, aim and scope of the presentation. Write down the aim in a well-crafted sentence and agree it with your customer (usually the principal member of the audience) and with others involved in planning and delivering the presentation. It is the highly distilled statement of requirement (SOR) to which you will be working.

Next decide the structure. You must not just start talking, go on until you have finished, and then stop. You must have at least three distinct parts, which you must flag clearly to the audience:

- Introduction: 'Say what you are going to say'
- Body: 'Say it'
- Conclusion: 'Say what you have said'

That is the very simplest case, perhaps for one person presenting on a single topic. In more complex cases, the body of the presentation will have several themes, each perhaps with its own presenter. That can be good; it gives variety of style, not least of the human voice. It can also be bad; it may disrupt the flow and lose valuable time. In these more complex cases, you will need a structure more akin to that of a management report, as discussed in Section 4.4 and Figure 4.10. In every case, though, the introduction is vital for explaining and launching the presentation, while the conclusion is vital for drawing the threads together at the end and recapping the main points which you want your audience to remember. Here are some other points to remember:

- During the actual presentation the visual channel is the main one. This is not confined to any visual aids you may use. It includes the appearance of the venue and the presenters, and the nonverbal communication they use. Feedback from the audience may be very direct and immediate, especially if they don't like what you tell them. That is one of the special challenges of a presentation.
- Your presentation may be in support of a management report. This is an important case because your audience may include the senior managers who have the authority to make decisions arising from your report but who may not have time to study your report in detail. You must get the key message across, as in the executive summary of the report.
- In that situation the scope and structure of the presentation need not necessarily be the same as the report itself. The presentation must complement, or support, the report. Generally, you will omit the noncontentious factual detail (data). You will focus on analysis (information and knowledge), recommended decisions and how to put them into action.
- You are not giving a lecture. A lecture has been defined as a method of transferring information from the notes of the lecturer to the notes of the audience without passing through the mind of either. That is not what you are aiming for!

If you are uncertain about structuring a complex presentation, use the ideas about management reports from Chapter 4. The rest of this chapter focuses on the planning and delivery of a presentation for which you have already decided the purpose, aim, scope and structure (PASS).

5.3 Planning a presentation

Planning a presentation, like writing a management report, is a project. There is a deliverable and a deadline. You need to plan. Time spent planning is saved many times over, by avoiding false starts and unnecessary mistakes. A controlled process of planning is more likely to lead to the delivery of a quality presentation at the required time.

The first thing to decide is who should attend. This should be apparent from considering the purpose, aim and scope which you have already decided. Include people who have a direct need to know, for the purpose of completing the management learning cycle (do-watch-think-try, as shown in Figure 2.8). Exclude hangers-on who may be interested but who have no direct role in taking things forward. You want a smaller audience rather than a larger one. This will create a more direct and intimate atmosphere. Focus on the real issues and you will stand more chance of being effective. Usually you want to stimulate an interactive exchange, not deliver a formal address.

The size of the audience will determine the type of venue. Sometimes you can choose the location. This can be important, because people are most confident and at-ease on their familiar, home ground. On the other hand they then tend to work on other matters until the last minute, arrive in the nick of time (or late) and are anxious to get away to resume the daily rush. All this means that their concentration is reduced. If you have a choice of location, consider these and the following points:

— For a presentation to an external customer, it is quite important whether it is held on your premises or theirs. If you are one of several firms bidding competitively for a contract, then the customer must have the choice. Sometimes they will want, on their own premises, presentations from several contractors in turn. Sometimes they will wish to visit your premises (and those of your competitors) to form a view about what sort of company you are. In that case, it is convenient to have the presentation in that location. You are seeking the business; you must give the customer the choice. Conversely, if you are seeking subcontractors for your work, make sure you exercise that privilege fully!

— Presentations which require members of the audience to come from afar may mean you get fuller attention once they arrive. However, they will expect to 'make a day of it' (or at least half a day). Besides the presentation itself, you will be involved in visit programmes, arranging transport and accommodation, and providing meals and other hospitality. This makes the whole exercise more complex to plan and more costly in your time and the firm's money. It needs careful thought.

— Mounting a presentation away from your home base can also be complex. You have to think about the supporting facilities like visual aid equipment. Do you rely on your hosts providing such equipment, or do you take your own? Are you able to rehearse properly? Something is bound to be different in the strange location. Worse still, there may be something you had not thought of: the leads won't reach, the slide cassette won't fit (that's quite likely), or the computers aren't compatible (that's guaranteed). At the very least a detailed liaison visit beforehand is needed. Try to persuade your hosts to allow you a private rehearsal in the venue to be used for the final presentation. Make sure all the equipment used at the rehearsal is the same as that which will be used on the big day.

Thus geographical location is one aspect of choosing the venue. The other is the choice of room. You are mounting a business presentation, not promoting an entertainment spectacle. Consider not only the presentation itself but also what you want to happen afterwards. It is no good delivering a brilliant presentation if you have not prepared the audience to pick up the threads and continue with moving the business forward. You must allow questions of clarification, but usually your aim is to move into a discussion about the issues which your presentation has raised, and to make some decisions about the next set of actions which is necessary.

For this purpose a smaller room is better than a larger room in relation to the size of the group. It should not be uncomfortably crammed, but nor should people feel that only half the intended audience has turned up. The more intimate the environment, the closer you can get to the audience (both physically and mentally), and the more likely you are to have a free discussion.

Think about the arrangement of the room. If you have to serve refreshments, keep all that near the door. Have the focus of the presentation (the dais, table, projection screens) away from the door. Many rooms are planned that way. Sometimes the doors are on one or both sides of the focus of the presentation. That means that latecomers provide a distraction to the audience. This in turn means that the latecomers are embarrassed. In that situation decide whether you should bar latecomers from entry until at least the formal part of the presentation is over.

Make sure that all members of the audience can see the focus of the presentation. They must be able to see the principal speakers and the screens or tables which will display any supporting material. This might mean it is better to have the focus of presentation on the narrow side of the room rather than the long side. Don't just check that you can see from every seat in an empty room; will people be able to see over or around others when the room is full? For a complex presentation think about the practicalities of speakers exchanging position. If they all have to go to a main lectern, can this be done smoothly? If there is a public-address or audio-visual recording system, is there going to be a problem with microphones or lights? Is anyone in the audience partially sighted or deaf? If so, consider the seating plan and the possible use of the public address or audio loop system to make sure they are not placed at a disadvantage.

There are a myriad of practical issues like these which affect your planning of the venue. It is very easy to forget something which could have a major adverse impact on the success of your presentation. An example is strong lighting, which on the day can dazzle speakers who are not used to it. There are two things to do:

- In your planning think through in sequence the whole series of events before, during and after the presentation itself. Ask yourself what could possibly go wrong. Include power cuts, fire alarms and blown projector bulbs in that.
- Insist on doing at least one rehearsal under conditions which are as near to those of the final event as possible. Try to have an audience for that rehearsal, not just stooges, but people who can make some friendly and constructive comment on your performance. Remember that people generate 300 watts of heat; the audience can soon change a pleasantly cool venue into an unpleasantly hot one. This will not help you or the audience to give of their best.

Consider the time of day that the presentation should be held. Sometimes you have no choice about this. Sometimes it is dictated by travel and other practical arrangements. However, if you do have the choice, consider these factors:

- Plan for the total event: presentation plus discussion/decision. Allow a generous estimate for the total time. Then plan it in relation to lunchtime.
- Most people are at their most alert midmorning, and suffer a relapse of attention and energy in the early afternoon. Ideally you should present straight after midmorning coffee, allowing enough time to complete the agenda before the onset of hunger and minds thinking yearningly about lunch.
- If you volunteer to mount your presentation straight after the audience has had a heavy lunch then you deserve what you get. However, if this is thrust upon you then try to recover the situation by building engagingly upon the rapport which you should by then have established. You will have an even tougher time than usual to maintain their full attention.

There are a myriad of practical issues like these to consider. Often it is fairly simple, but sometimes it is very complex. If you need further advice, see Seekings (1989).

Now assume that you have now decided the audience, the venue (geography and room) and the start time and duration of your presentation. You can turn at last to the planning of the presentation itself.

Will it be a team effort, with several presenters? For short and simple presentations (perhaps up to 10 minutes) a single presenter is best. If you have decided that something more complex than the basic three-part structure is needed, then there can be advantages in having several speakers. But consider these points:

— Always have the same person (the lead presenter) to present both the introduction and the conclusion. As with a management report, those two elements should be a coherent summary of the whole performance. Having a single presenter for them emphasises that. It uses the time when the audience's level of attention is at its greatest. For the first few minutes you are sure to get their attention. You will get a high level of attention in the last few minutes only if they know that the presentation is drawing to a close. Your reappearance signals this to them, and you should gently remind them of what they have to do next. This might be to seek clarification or amplification of what has been said and then lead into a period of discussion. If you are not yourself the chair for the whole session, then liaise closely with the person who is. Otherwise you will not achieve a smooth transition.

— The audience's natural attention curve shows a slow drop for about ten minutes, then a rapid drop to a low point about 30 minutes into a presentation. It rises to a peak in the last five minutes, provided you tell them that the end is nigh. You will use the natural high points for the introduction and conclusion, as mentioned. You must think carefully about the remaining period, the body of the presentation, when the audience's attention may wane.

— Consider using different speakers, with various personal styles, in the body of the presentation. Each time a new speaker appears, there is an upward blip of attention. Each individual should try to build on that.

— Your presentation will have some natural highlights and some more prosaic parts. Try to vary the texture of the presentation, and the use of visual aids, to achieve variety. Often there are people who are obviously appropriate to present on particular topics. You may not be able to change that. However, as always you must think of your audience; they might view some people as boring presenters of boring topics. If these are necessary, at least keep them short and don't have too many of them in a row.

— There is nothing wrong with having a natural ebb and flow of interest during your presentation. You cannot expect people to be gasping with rapt attention for the entire proceedings. But at least try to manage things so that they are gasping with rapt attention during the important bits.

So much for the structure, and the role of individual presenters. Every presenter then faces a fundamental problem: how to know what to say during the presentation. There are six options:

1   Rely on the right words coming into your head at the time. Don't do this unless you are a natural orator, and you live and breath the subject on which you have to speak. For the rest of us, and certainly in business presentations, you must be more methodical. Otherwise you will probably freeze as your mind goes blank, or say something irrelevant, or get the timing wrong, or ruin the way your contribution fits in with the rest of the proceedings. No matter how confident you might feel beforehand, do not do this!

2   Write out a full script of what you have to say, and then read it out. This is safe, rather formal, and usually extremely boring for both you and the audience. Occasionally, however, it is necessary. Some presentations have to be delivered word-perfect, because they are 'on the record' (formally attributable) and on an important or complex topic. In this exceptional case, a written script has to be used. Reading a script is also the safest option for a beginner, or one who is racked by nerves about having to give a presentation. In that case, make sure that you try to write 'spoken English'. This means using everyday language rather than the more formal language of a management report. It means using a vigorous turn of phrase, as you would

in conversation. It means having short, simple sentences rather than long ones. Be direct. Give the facts. Draw your conclusions. When you deliver the text, give the inflection and gesture which you would in conversation. It is not easy to read a prepared script and make it sound fresh and sincere. But if you feel you have to read a full script, then you must try to make it come alive.

3   Write out the full script, learn it, then recite it from memory. This is not a good method. Even if you use notes to jog your memory, you will almost certainly get some passages wrong and omit others. Unless you are a skilled actor, it will be evident what you are doing and you will appear insincere. However it is sometimes useful to learn verbatim a few key sentences of your presentation. These might be the most important ones in the presentation: the first few and the last few. Learning the opening sentences gives you confidence in the first few seconds of your presentation. At the end it helps ensure that you finish gracefully and effectively rather than droning to an uncertain halt.

4   Write out a full script, but then speak from brief notes. This is perhaps the best method for the beginner. In writing and reading over the script, try to get the ideas into your head, rather than the precise sequence of words. At the time of the presentation you do not use the script. Instead the notes jog your memory about the ideas. You then put those ideas into words at the time. The words will then be more spontaneous, natural and conversational. They will vary slightly each time you give the presentation. But on each occasion the same ideas should have come across clearly.

5   Use structured notes, having thoroughly rehearsed the ideas and words mentally. This is a more advanced technique. You plan and structure what you have to say, and produce notes which reflect that structure. The notes may consist of headings or key words, or be in the form of a mind map. Mind maps are described by Buzan (1988 and 1989). On several occasions before the presentation you use your notes and mentally compose and deliver the speech. You can say it aloud or whisper it. Don't just think the words; you will trip up, and also get the timing wrong.

6   Use method 5, but do not use the notes when you actually deliver the presentation. Instead, have them in your pocket just in case your mind goes blank and you need to refer to them. This is the technique for the expert! Do not use it unless you have a high confidence that you will not need to refer to the notes. The audience will be fully aware that you have lost the thread of your thoughts, when you reach for your notes, fumble to find the right place and then resume speaking.

Which of those methods sounds right for you? How many have you used in practice? Once, as a schoolboy, I used method 1; I never did so again! Everyone has to develop their own technique. My advice is to start with method 4, while using method 3 to make sure you get the key opening and closing sentences right. Gradually you can develop towards method 5. One thing is certain: you will improve with practice. So, take every opportunity to contribute to presentations, even if you are naturally nervous about doing so. You will reap the benefit of increased confidence and improved technique.

Next in planning your presentation, consider visual aids. They should be exactly that: aids to your presentation. They should not be allowed to become the dominant aspect. Use them with caution: all the time the audience are looking at a visual aid, they are not looking at you. That means you cannot maintain eye contact, which is one of the most important methods of maintaining attention and getting feedback.

That said, a visual aid can be very helpful. Section 4.2 considered the use of figures and diagrams in a management report, especially to present quantitative data as meaningful information. The same benefits apply, and to a greater extent, in presentations. In a presentation, there is no opportunity to pore over a table of numbers; you have to give the audience the results of your analysis, and the most effective way to do that is often graphical. Modern IT helps us to produce high-quality visual aids quickly and conveniently. But this is itself a hazard; people produce endless diagrams of bullet points which they read from the screen. It is better for the speaker to have the notes, and maintain eye contact with the audience for a greater part of the time. When Lou Gerstner took over as chairman of IBM in 1994 he was subjected to a wave of corporate presentations, each with a sheaf of Vufoils for the overhead projector. Eventually he banned the Vufoils. The presenters then had to think more carefully about what they were saying, as opposed to what they were showing.

Here are some tips about different kinds of visual aid:

> *Blackboard and chalk:* Some of the fustier academic institutions still have blackboards. They are messy and inconvenient. Never choose to use one in business, but if it is thrust upon you then adapt the whiteboard technique as best you can.
>
> *Whiteboard and dry marker:* These are available in nearly every business conference room and most offices. There will usually be a wiper or duster. Before the presentation, make sure the board is clean, or contains only some detail you want (such as the title or the timetable). There should also be a stock of dry markers of various colours. It is most important to use only the correct kind of marker, otherwise you incur administrative wrath when people have to come and clean the board with white spirit. Test out the markers before you start. Often they are dried out, because people just put them back when they find they don't work. If you find they don't work, put them in the wastebin. Write a few words, of the size you will be using, in several colours. Go to the back of the room and see if you can read them. You will probably decide you must write larger, and avoid certain colours (especially red). The whiteboard is useful for building up complex diagrams during an interactive session with your audience. This is invaluable in teaching. It is less appropriate in a formal business presentation, but could help in any discussion or brainstorming session afterwards.

Flipchart and marker: This is a pad of paper, usually of A0 size (841 × 1189 mm). Being smaller than a whiteboard, it is less visible. However, you can have prepared items which you can reveal at the appropriate moment to the audience's surprise and delight. Remember to cover them up after the critical moment. Some chart devices will produce an A4-sized copy of what is written; this can be useful for the secretary, or to give to people as an instant record of a discussion. This is more useful for a meeting than a presentation.

Overhead projector (OHP): This is a ubiquitous and very flexible tool and is therefore grossly overused. Few people use it in the way originally intended: sitting beside it and writing directly onto acetate foil, as a substitute for writing onto a blackboard or whiteboard. That technique can be quite effective: the writing appears large, and you remain facing the audience. However, it is a strain on the eyes to write while the OHP is turned on. Most people, therefore, use prepared Vufoils (or acetates). You can have multilevel Vufoils which build up the detail as you work through the presentation. Make sure these don't get too complex, and that the detail is visible from the back of the room. Remember that words are not really visual aids, so don't have long lists of abstract nouns. Even if you plan a sequence of Vufoils, allow 2–3 minutes for each. They may be familiar to you, but they are new for the audience. There is a practical point about OHP machines. Some portable ones are very neat, but rely on reflection from a mirror underneath the foil. If the foil curls or wrinkles, you get a double image. A cardboard frame, with the Vufoil fixed underneath (rather than on top) helps with this. Be careful about the size of frames. Most Vufoils now are prepared in A4 (297 × 210 mm) format, but many machines will only project a smaller size. It is safer to assume that the working area is 245 × 200 mm; these are the inner dimensions of the older cardboard frames.

Slide projector: It is now easy to produce high-quality 35 mm slides. For photographs they are very effective. However, there are several disadvantages with using slides. The room usually has to be darkened for this part of the presentation; this can be slightly disruptive, and make it difficult for people to take notes. It also encourages slumber, so don't use slides after lunch. Nowadays, colour photographs can be transferred to an OHP transparency. This can then be integrated into a series of other Vufoils. This reduces disruption and retains the advantage of the photograph. So, avoid slides unless there is a set of photographs which would be unwieldy on the OHP.

Video: Modern conference rooms have this facility built-in. A three-colour projector is directed at a fixed screen. You must not disturb the alignment. Most projectors revert to standby after a while; make sure you have the infrared control and know how to reactivate it before you want to show your video. The alternative is a large-screen monitor, or several in the case of a big room. Again, monitors tend to revert to standby when you don't expect it. Finally the video player itself has to be operated. They all have buttons in different places, so you might ask someone to assist you. It can make an entertaining spectacle (for the audience) when you are trying to zap the projector or monitors into life and start the tape moving. Make sure you check all this before the audience arrives. Make sure that the player is the same standard as your tape. This is usually VHS/PAL in Europe; it may be BetaMax/NTSC elsewhere. Despite these practical hurdles, video can be very useful. Keep your clips short: up to four minutes. They must not take over from your presentation. And make sure the material fits in well and is appropriate to the level of your audience.

Physical objects: These can be very effective if produced from hiding at the critical moment in your presentation. 'And here is the very thing. . .' is sure to get the audience's attention. If you can, put it away after the relevant portion of the presentation. Certainly do not pass it round the audience. That is disruptive and most people get it to look at when you are trying to engage their attention on something else. You will then lose out to the fascination of the object itself.

Working models: The same principle applies. Don't have the apparatus there as a tantalising puzzle until you use it. Unveil it; use it; cover it up. Make sure it works. Demonstrations that fail are a disaster. A good rule is 'Never give demonstrations in public'.

Thus there are many ways of providing visual aids. You must be discerning. Ask yourself 'Is it visual?' and 'Is it an aid?'. Only if you answer yes to both questions should you proceed. You should also remember that visual aids are no use unless they are visible and legible. Most OHP and slide screens, and all video projectors and monitors, are arranged with the longer dimension horizontal. This is called landscape format, as opposed to portrait format which has the longer dimension vertical. The ratio of the longer side to the shorter side is different for the various media. It is 1.50 for 35 mm slides, 1.41 for A-series paper-sizes, 1.33 for video/TV, and 1.78 for the emerging high definition TV (HDTV) standard. Remember that these differences can affect the appearance of your diagram when it is transferred from one medium to another. In practice it is a good idea to produce diagrams in the 1.41 landscape format. This means the diagrams will fit neatly into a written report and will also make visual aids for supporting presentations.

Looking at that more closely, for the international paper sizes the ratio of the sides is equal to the square root of two (i.e. 1.41). This means that when you cut the paper in half the proportion is unchanged. The basic size is A0 (1189 × 841 mm), which has an area of one square metre. A1 is half of A0, and so on. This makes up the A-series of paper-sizes. There are intermediate sizes: the B-series for posters, and the C-series for envelopes. The useful sizes for our purpose are listed in Table 5.1. Most reports and the handouts for presentations will be in A4 portrait format. Reports may have fold-out pages in A3 landscape format. If you use the C-series for the sizes of diagrams they will fit neatly onto the paper in each case. For example, you can get two C6 landscape diagrams, or eight C8 landscape diagrams, onto an A4 sheet. In the A4 portrait format (for a report

Table 5.1   Paper and diagram sizes

Designation	Use	Long side (mm)	Short side (mm)
A3	paper (fold-out)	420	297
C4	diagram: 1 on A3	324	229
A4	standard paper size	297	210
C5	diagram: 1 on A4	229	162
A5	small paper size	210	148
C6	diagram: 1 on A5; 2 on A4	162	114
C7	diagram: 2 on A5; 4 on A4	114	81
C8	diagram: 4 on A5; 8 on A4	81	57

or handout), this means that both text and diagrams are presented conveniently: the text in portrait format, and the diagrams in landscape format. Head-twisting by the reader is avoided. Modern PC packages enable you to change the size easily, keeping the same 1.41 ratio of sides. Sometimes, of course, you will integrate diagrams with text.

What about legibility? It is a common error to make the lettering too small. This also tempts us to include too much detail. This is often because we sketch out the diagram on an A4 sheet of paper (or a PC screen of similar size) about 50 cm in front of us. The long side of the A4 sheet is about 30 cm, so it subtends an angle of about 35° at the eye. However, in the presentation room the diagram may be projected onto a screen 2 metres wide, and some people may 15 or 20 metres from it. For them the diagram may subtend an angle of only 12°. To make the image legible for them, the A4 sheet must be legible to you from a distance of 3 metres. If you are giving a presentation in a large room it is worth doing some calculations of that sort. However, as a rule of thumb any diagram should be clearly legible from a distance ten times its longer side.

So try reading an A4 sheet of typescript from 3 metres away. It's difficult. So don't make OHP Vufoils by direct copying from books or reports; enlarge or redraw them first. Now try looking at Figure 5.1 from a distance of ten times its width (1.7 metres, as it appears in the book).

Here are some conclusions you might draw:

- The capital letter height of any writing should be at least one-twelfth of the height of the diagram.
- This means you can only get eight lines of text onto the diagram. Of course, you should not be using so much text on a diagram, anyway!
- Bold, simple letter-styles are better than thinner, more exotic ones.
- Black and white is most legible. Colour can be very effective in moderation, but remember that 6% of your audience may be colour blind.
- A bold title is welcome. Put it at the top for a presentation diagram (unlike a diagram in a report or book). A reference number is useful.

This text is much too small

This is the smallest practical size

This is more comfortable

Useful for headings

And bigger headings

Better for visual aids

With this for headings

THE MOVING FINGER WRITES; AND, HAVING WRIT,
MOVES ON: NOR ALL YOUR PIETY NOR WIT
Shall lure it back to cancel half a Line,
Nor all your Tears wash out a Word of it.
Rubaiyat of Omar Khayyam (2nd Ed,1868), Quatrain 76

Figure 5.1   Legibility

● Using block capital letters reduces legibility. Fonts with serifs are clearer than sans-serif fonts.

Visual aids, and how they relate to reports and handouts, are important in business because presentations are often given in direct support of management reports. You can save yourself a lot of time and help the audience if you have a systematic approach of the kind suggested. There are some related issues about handouts at the presentation. If you decide to have a handout, you have several choices:

1   Issue the handout at the start. This is simplest, as it can be given to people on arrival or placed on the chairs. The trouble is they will start looking at it before you start. You lose the impact of your opening remarks, and any surprises you may have planned.
2   Issue the handout at some point after your introduction. This means you have their full attention for the introduction, but the same dangers thereafter. It is also disruptive to distribute the handouts.
3   Issue the handout at the end. This maximises their attention to the presentation. Tell them there will be a handout; they will then know they need not take comprehensive notes. The disadvantage is that they cannot annotate their copy of the handout with additional points during the presentation.

Some handouts comprise only copies of Vufoils. This is unhelpful. At least you need a cover page giving the date, place and topic of the presentation, and the names and organisations of those involved. Remember that if you have used Vufoils properly they are unlikely, by themselves, to give a proper summary of the presentation. You could produce bullet-point text notes which refer to the attached or adjacent copies of diagrams. If you have planned your presentation well it should be simple to produce such a handout. It will be much more useful to your audience. However, do not issue a full-text document (or even the related report) as a handout at the time of the presentation. People will rustle pages, get diverted, and fumble to find the diagrams they see on the screen. Keep it simple, for the benefit of all concerned.

So you have now decided on the audience, the venue (location and room), and the timing. You have decided who will present each part. Presenters have made their speaking notes and rehearsed their own contributions. The visual aids and handouts have been planned and prepared. You have thought through the whole sequence of events and pondered what could go wrong.

It is time for a rehearsal. In fact, it is time for more than one rehearsal because in all except the simplest case you will need several. If you are co-ordinating a complex presentation, you may wish to go through each person's contribution with them individually. Quite soon, though, the whole team should assemble for the first rehearsal. This should be viewed as a stagger through. Don't worry too much about the timing at this stage. Concentrate on getting the logistics right: people moving about smoothly and the unobtrusive use of the visual aid equipment. At the second rehearsal you should do better, and you can focus on timing and the finer points of style. This is where you might ask some independent person to sit in and give helpful comment. You should always have a full dress rehearsal where you hone it into a polished and professional performance. Unless it is utterly impossible, the final rehearsal (at least) should be held in the actual venue. Even contractors inviting you to present on their premises are usually sympathetic to that. It shows that you intend to do a professional job.

By this time your team should be well motivated and reaching a peak of performance. Don't over-rehearse; it can make the final product wooden. Time things to keep the adrenalin active, whilst achieving an underlying confidence based on the detailed planning you have done.

5.4 Delivering the presentation

As with most things in life, the planning and practice for a presentation are the major factors in its success. That's why the major part of this chapter has been about planning. On the actual day of the presentation all should go smoothly.

There are some details that are important on the day and which will probably have been resolved at the time of the final rehearsal. Here are some:

— Decide what you will wear. If it is a team presentation, should you try to harmonise your dress? Dressing smartly tells the audience you take the occasion seriously. Remember not to put your hands in your pockets. This can be seen as too informal, and you may be tempted to rattle coins or keys. Avoid any personal or verbal mannerism which could detract from the effectiveness of your message. Silence bleepers and mobile phones.

— Arrive at the venue in good time. Each presenter should make sure they have their own notes, and visual aids (unless these are all being handled together).

— Make contact with the adminstrative people in charge of the event. Check out the audio-visual equipment and lighting. Agree who will control what. Make sure you know the relevant emergency procedures. Check out all the visual aids; in the case of video or audio tape, bring a test tape so that you do not lose the place on the tape you wish to show during the presentation. Check about reserve equipment and spare bulbs. Have a pointer available, and a gavel for the chair. Provide water for the speakers. Make sure the temperature is on the cool side; it will warm up when the room is full.

— Complete these preparations before the first members of the audience are due to arrive. Then relax, welcome them, and join them for coffee. After all the preparation it is a good thing to take your mind off the presentation before you actually deliver it. Don't sit in a side room biting your fingernails.

— At the appointed time the chair should call the group to order and introduce the proceedings. This may be your job as the presenter. In any event make sure that your name and the names of your team are announced either at the start or during your own introduction.

— Few presenters are not feeling a bit nervous at this stage. That's not a bad thing. If you do not feel nervous before a presentation, and exhilarated (or even exhausted) afterwards, you have not done your best. Breathe deeply. Quietly sucking a mint can help prevent voice problems, but make sure you finish it before you have to perform. A sip of water also helps.

— The delivery of the presentation will, of course, have been thoroughly planned. Once it's under way this is often the easiest part of the proceedings. However, you do need to be alert to two things. One is failure of equipment; you then have to decide whether to stop to deal with the problem or carry on without. If the audience have a handout, this can save the situation if the OHP fails. The second thing you need to watch is for feedback from the audience. If you are using the recommended method of speaking afresh from notes of key ideas, then you have some scope for varying your style of delivery. If the audience look puzzled you may need to spend more time in explanation, and omit some of the detail. If they look bored you should perhaps adopt a more lively style. Don't overdo that, or act outside your own character.

— Show enthusiasm for your subject. It has been said that the most convincing people show messianic enthusiasm, tempered with reality and the healed scars of experience.

— You will even consider humour. This is a dangerous area, as mentioned in Section 3.3 in the context of speeches. You should tell a joke only if you are totally confident that it is appropriate to the audience and the occasion, will not cause offence to anyone present, and is likely to be perceived as very funny. That rules out most jokes. It is better to have lighter forms of humour like understatement or gentle jokes at your own expense. Laugh at yourself, but never laugh at your subject or at your audience.

— If possible, questions should be kept until the end. Say that at the start. If, nevertheless, someone interrupts with a question, either answer it quickly or politely defer it until the end. In the rare event of someone getting agitated or angry, try to defuse the situation with a polite and noncontentious deferral. But in that case make a note to ensure that you do address the issue in question time. Otherwise you will have a very disgruntled customer. However, most people if asked directly and politely will wait until the question period. Even your boss, and your boss's boss.

— Your presentation will end, of course, with your conclusion which draws together the threads, summarises the key points and indicates what will happen next. Then say 'Thank you' and sit down. Even in a business situation there can sometimes be applause. If this happens, acknowledge it gracefully and indicate that it has all been a team effort. More likely, though, you will go straight into questions.

— Questions should be firmly under the control of the chair. This may be you as the presenter or another person. It is a good idea to have all the presenters available on chairs to one side, ready to answer detailed questions. Involve them anyway, even if you could answer the question yourself. You might add a comment at the end; this will then have more weight than if you had answered the whole question yourself.

— Invite questions of fact or clarification first. Then move into the more subjective areas. At this point, the emphasis shifts from the presentation to a business meeting. The methods for handling meetings are discussed in Chapter 6.

During the delivery, bear in mind that the more you say the less people will remember. This has four implications:

1 If you have to itemise a list, keep it short. Try to have three items. Subconsciously we expect a list to have three items, and audience behaviour reflects that. It is called the rule of three.

2 Vary the tone of your voice and the pace of the delivery. Pause, smile, and repeat. Emphasise the main points.

3 Remember the power of silence. It has been said that the most beautiful sounds are heard on the edges of silence. That applies to music; it also applies to presentations.

4 Brevity is a virtue. The Gettysburg Address took two minutes to deliver. The main speech on that occasion lasted two hours. Which do we remember now?

That completes your presentation, almost. When the audience have dispersed and you and your team are alone you will hopefully share a sense of achievement. Even if your presentation was the prelude to some unhappy decisions, at least you did your job professionally. So, be ready to indulge your sense of relief and achievement. You might enjoy a drink or a meal together, even if it is just sharing a table in the staff canteen.

Eventually, when the euphoria has worn off, you should take stock of what happened in a review. Get together for an hour to identify what you learnt from the whole experience. This will enhance your individual and corporate learning, and mean that you will do even better next time. Remember it is the learning individuals, and the learning organisations, which prosper in today's business environment.

5.5 Summary

This chapter has been the first of three on special aspects of communication in business. Presentations are an important method of commmunication, both internally (with colleagues) and externally (with customers and others outside your department). They can be needed for many reasons, but notably to convey information, or to support or promote a product. This product may be a good or a service, internal or external to the company. An important case is a presentation in support of a management report which is being considered by senior people.

For every presentation you must decide the purpose, aim and scope (as discussed in Section 2.4). You can then decide the structure. This should have at least three parts: introduction, body and conclusion. For more complex presentations, use the same principles as for structuring a management report. This completes the overall planning: purpose, aim, scope and structure (PASS).

Section 5.3 turned to the detailed planning of the presentation. The first step is to decide on who should attend. You should aim for a small audience comprising the people who are really involved in carrying the matter forward. Then think about location. There are advantages and disadvantages for both you and the audience for 'home' and 'away' presentations. As presenter, you may well prefer a home event, but always give your customer the choice. Consider the other aspect of venue: the room itself. This should be as small as possible without giving discomfort. You want to encourage interaction in a relaxed and comfortable situation. Think about the arrangement of the room, especially the relative positions of the audience, the focus of the presentation, and the doors. Remember the effects of lighting and audio systems. Consider what could go wrong, and then arrange a full rehearsal in conditions as near as possible to those of the final event.

Consider the total agenda for the event. It is likely to be more than just delivering and receiving a presentation. The most important part is likely to be the period of questions, discussion and decision. Allow plenty of time for all that, then plan the timing in relation to lunch. Choose the morning in preference to the afternoon.

In planning the presentation itself, think again about your structure. If it is simple have a single presenter. If it is complex make sure the lead presenter gives both the introduction and the conclusion. The latter should alert the audience to the fact that the presentation is drawing to a close, and remind them of what will happen next. In between the introduction and the conclusion, you have to battle with the lower attention level of your audience. Vary the texture of speakers, styles, topics and visual aids to help achieve an event with a rhythm which ends on a high note.

There are six methods of knowing what to say. Beginners should prepare a full text, but then speak from notes which remind them of the key ideas. It sometimes helps to learn the important opening and closing sentences verbatim. Aim to develop your technique. Eventually you will use the notes to remind you of the key ideas, and you will express those ideas in speech more spontaneously. You need to practise that several times beforehand. Take every opportunity to develop your technique.

Visual aids must be exactly that. In business, the OHP, flip-chart and whiteboard will suffice for most presentations and the ensuing discussion. Short video clips can be very effective, but be wary of using slides. Plan the use of visual aid equipment carefully, and make sure the resulting images are clearly visible to all members of the audience.

It is good policy to produce diagrams in landscape format, with the longer side horizontal. If you keep the ratio of the longer to the shorter side at 1.41 the diagram will be ideal for fitting into reports and handouts. The C-series of sizes is good for that. Make sure the diagram is clearly legible from a distance of ten times its longer side. Use lettering whose capitals are at least one-twelfth the height of the diagram; this implies no more than eight lines of text. Use a bold title, at the top. If you are giving handouts, use key diagrams but have a title page and brief notes as well. Decide when to issue the handout.

You need a rehearsal. For a team presentation you need several rehearsals. The first is a stagger through; the last is a full dress rehearsal in the final venue. Don't over-rehearse; keep up the team motivation and expectancy.

Section 5.4 considered the actual delivery of the presentation. This includes preparation, reception of the members of the audience, the opening of the proceedings, the delivery of the presentation, questions and the smooth flow into any wider discussion. Afterwards, build on the sense of achievement which you and your team now have. Later, get together for a review of the whole experience, to maximise individual and corporate learning from it.

As with management reports in the preceding chapter, the most complex and demanding type of business presentation has been considered. Many will be much simpler than that. However, you can apply the same principles and use the ideas as a checklist to make sure that even a small and informal presentation is done professionally and effectively.

Finally, it is the visual channel which is most important. People will decide whether they trust you and what you say largely by what they see. Be brief, and remember the power of silence to achieve emphasis. The Revd Sydney Smith (1771–1845) said of Lord Macaulay 'He has occasional flashes of silence which make his conversation perfectly delightful'.

And remember to get silence before you start to speak. Then say something which will engage the audience's attention. A speaker from Africa to a World Health Organisation conference started his presentation by saying in a deep, slow voice 'My father once ate a man'. The audience listened.

Chapter 6
Business meetings

'Meetings are indispensable when you don't want to do anything'
J K Galbraith (1908–)
'An extraordinary affair. I gave them their orders
and they seemed to just want to sit round and discuss them'
1st Duke of Wellington (1769–1852),
on leaving his first Cabinet meeting as Prime Minister

6.1 Introduction

You recall that Chapter 2 established the foundation for the detailed chapters of this book. You may wish to remind yourself from Section 2.4 about the management learning cycle (Figure 2.8; do–watch–think–try) and the information value chain (Figure 2.9; data–information–decision–action). How do business meetings accord with those ideas?

Meetings, like the reports and presentations we have considered in the last two chapters, come in a wide and varied range. Here are three types of meeting:

● One-to-one encounters, whether planned or spontaneous. In this case we are really engaged in a purposeful conversation. We need the verbal and nonverbal skills detailed in Sections 2.2 and 3.2. Section 3.3 mentioned the special cases of negotiation and interviews.
● Small groups assembled deliberately or spontaneously in the operational environment of the business. They may be getting data about a problem, discussing how to solve it, or simply 'seeing and being seen'.

• Groups convened for some particular purpose, away from the operational environment of the business. The purpose is to 'think', individually and collectively, and arrive at decisions. The input to the meeting is information; the output is a series of agreed decisions which can be put into action.

It is the third type of meeting which we are concerned with in this chapter. First consider the purpose of such meetings in a little more detail. Then consider how to plan the meeting, and finally, about the running of the meeting itself, including the topic of group development. This approach echoes that used in Chapter 5 to think about presentations. That is no coincidence; the two are closely related. But the detail is different.

6.2 The purpose of meetings

Meetings must be purposeful. Everyone who attends should be clear why they are there. Anyone who is not needed for the achievement of the purpose should not be there (except in the special case of giving training or widening experience). Everything that is done should contribute towards achieving the aim of the meeting.

The meeting is an example of business communication between several people in a group. You therefore need to define the purpose, aim and scope of that communication, in the way discussed in Section 2.4. It is likely that the meeting is concerned with the middle part of the information value chain, as shown in Figure 6.1. The diagram is an annotated version of Figure 2.9. For each major item, shown as a shaded area, the meeting may

1   take in processed data, information and knowledge about the situation
2   think about and discuss the situation, to deal with any identified problems
3   reach a decision, or a set of decisions
4   plan the implementation of those decisions

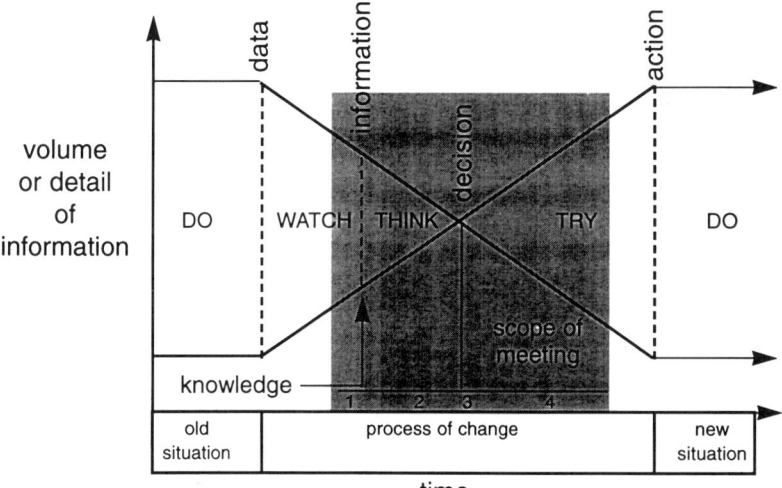

Figure 6.1   Scope of meeting

These four tasks are shown in sequence at the bottom of the shaded area. The balance of time and detail between the four tasks will vary. Some meetings just do tasks 1-3; planning the detailed implementation of those decisions is left for other groups or occasions. Those lucky people might themselves have a separate meeting, wholly concerned with task 4. So the flavour of meetings can vary. By taking the general case (all four tasks) you will have a template to deal with most of them.

You have now decided the purpose, aim and scope of the meeting. You must next decide the structure (and thus complete the acronym PASS). Do you just follow the shape inside the box in Figure 6.1? Yes? No? What has your experience been? Have you ever seen people in meetings

- not sure what was going on?
- asked to provide information, but not warned about this beforehand?
- fail to seek agreed decisions?
- fail to reach agreed decisions?
- try to push their own view too hard?
- impose a decision from the chair?
- take decisions which could not be put into practical effect?
- take decisions, but fail to agree how they would be put into effect?

These symptoms are horribly common. They reflect poor planning and running of meetings. You must resolve that you will do it better than that. The shape inside the box in Figure 6.1 is converge–diverge. If you just try to follow that, there will be trouble.

You should start your planning by considering the aim of the meeting again. Write it down. Refine it. Discuss it with those likely to be involved in the meeting itself, or who have an interest in its outcome. Get it right. That is then the aim which should be in the mind of everyone when the serious business of the meeting starts. At the start of the meeting the chairperson (whom I shall call 'chair' from now on) should make a short introduction, remind people of the aim, and indicate the structure of business thereafter. For each major item of business the converge–diverge process shown in Figure 6.1 must be similarly introduced and summarised at the end. The shape of each major agenda item is therefore as shown in Figure 6.2.

If you have studied Chapter 4 on report writing, you should compare this diagram with Figure 4.11, which showed the shape of an example management report. There are some close similarities, so the principles discussed there will apply here too.

The meeting will usually have several such major agenda items, to achieve the overall aim. The shape of the whole meeting might then be as shown in Figure 6.3. It looks pretty complicated. It has to be, if you are to achieve useful business results. However, several things stem from this.

— The business of a meeting is complex, but a diagram like Figure 6.3 can help you to plan and run it effectively.

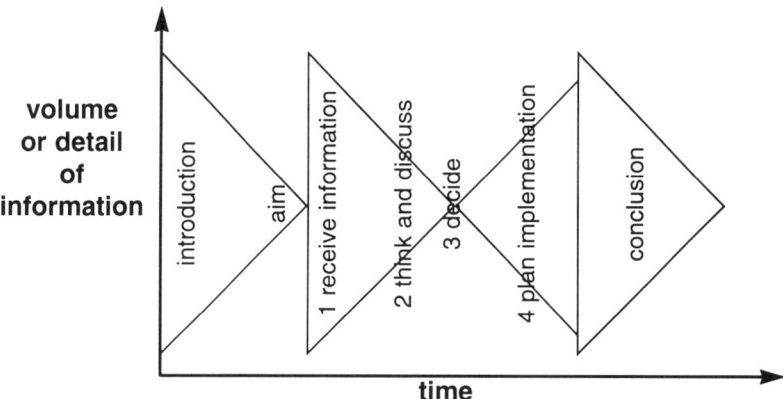

Figure 6.2 Shape of agenda item

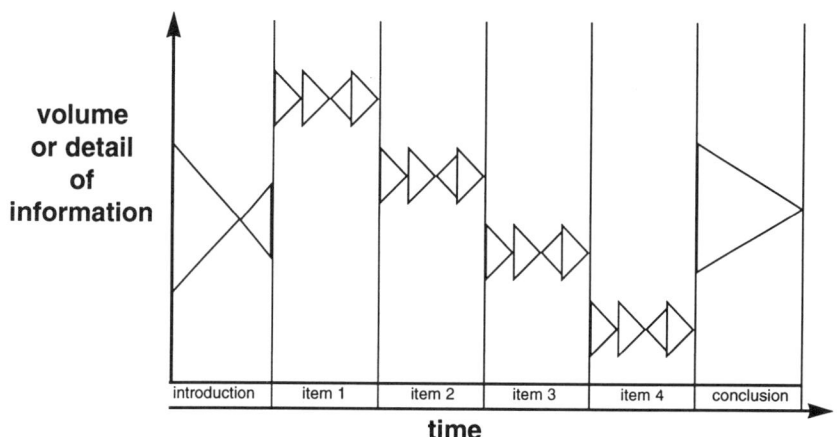

Figure 6.3 Shape of meeting

— The chair has a vital role in managing the proceedings. They must introduce and sum up the whole meeting, and introduce and sum up each major item within it.
— The aim and the plan will together determine who should attend. Equally important, they will determine who should not attend.
— Those attending will need an agenda beforehand. Besides administrative details, this should tell them the aim of the meeting and the structure of agenda items. If they don't receive an agenda before the meeting, they will not be well placed to contribute effectively to the meeting. They may not even come.
— For each major item the agenda should make it clear how the starting-point information will be provided. Occasionally, people will bring it in their own heads. This might be appropriate at the early stages of dealing with a problem,

when people are still trying to agree what the problem is. More often, there must be some formal way of providing information for the meeting, such as

- Asking an appropriate person to assemble the information and then present it verbally at the meeting. This is fine for a fairly simple matter. For a more complex matter, a short presentation would be better.
- Asking someone to prepare a briefing paper which can be circulated with the agenda before the meeting. A briefing paper is simply a management report whose scope is to provide information.
- A combination of briefing paper and presentation. This is the most effective method. For those (the majority, one hopes) who have read the paper beforehand it provides a useful reminder of the facts and perhaps a fresh perspective on them. For those (the minority, one hopes) who have been unable to read the paper beforehand, it ensures that they have at least some information on which to base their input to a sensible discussion.

To help or speed up the discussion of a major item, a more comprehensive management report might be prepared. This would have greater scope than a briefing paper; it would have analysis, conclusions and recommendations. It would be circulated in good time before the meeting, preferably with the agenda. At the meeting the report might be endorsed as it stood ('rubber-stamped'), or used as a springboard from which the meeting can reach its own conclusions. Whatever happens, the process is helped if the management report meets the last two criteria identified in Section 4.3: it should provide multiple levels of detail; and it should separate fact, analysis and opinion.

The technique of circulating briefing papers or management reports which support individual items is quite common for senior committees. Often the senior people will get their own staff to examine the papers and prepare a brief for them before they attend the meeting. The whole process can become slow and bureaucratic. One reaction to that is to insist that all papers comprise only one sheet of A4 paper. This certainly focuses the mind of those who have to write them, but for complex issues there is a grave danger that important detail is lost. It might be better to have a report written in accordance with the principles of Chapter 4. It should have a good executive summary, an introduction and conclusion which make sense when read together, and all the supporting detail in appendices so that the body of the report can be kept short. If you find yourself charged with writing papers for consideration by senior committees, take the job seriously. You can have a very great influence on what happens.

You have now got the purpose, aim, scope and structure (PASS) for your meeting. You can turn your attention to the practical details of planning and running it.

6.3 Planning a meeting

For a meeting of a formally constituted body like a committee, the detailed planning work usually falls to the secretary. However, the secretary should consult closely with the chair on important topics such as the agenda and the venue. For a less formal meeting, the person who is convening it often acts as chair and does the preliminary planning. In either case, the practical tips given here will be of use.

The agenda is the most important thing. Its main items should emerge clearly from the prior planning work described in Section 6.2. You will need to add details of the venue and the times. For a complex meeting it is helpful to give approximate times for the major agenda items. This helps people who have to come for only some of those items. It also shows the overall plan to which the chair will be working. If it is necessary to shorten discussion on a particular item, this can be done less contentiously if it can be seen by everyone that the chair is only trying to keep to the previously published plan.

A formal meeting will usually have an agenda like this:

1   *Welcome/introduction:* The chair will extend a welcome to those present, make introductions (possibly going round the table in turn), remind everyone of the aim of the meeting and the broad plan reflected in the agenda. The chair will also mention timing, and any other necessary administrative information.
2   *Apologies for absence:* These should usually be notified to the secretary in advance. At the meeting the secretary will report these, as a number or by individual names, and ask if there are any more. With a formally consituted body it is important to check that those actually present make up a quorum. A quorum is the minimum number or combination of members empowered to conduct business. An inquorate committee cannot take decisions.
3   *Minutes of the last meeting:* These may be circulated with the agenda. It is then assumed that recipients have read them. Under this agenda item they can request changes for the purpose of making the record accurate. They cannot request changes to include what they wish they had said or to change what was agreed. Sometimes the minutes have not been circulated, for practical reasons. This often applies to public and voluntary bodies. In that case the secretary will have to read out the minutes so that they can be considered by those who were present at the earlier meeting. This is an unsatisfactory procedure, but often unavoidable. In any event the chair will usually invite a proposer and seconder that the minutes be accepted.
4   *Matters arising:* This gives opportunity to discuss or report progress on items which were minuted from the previous meeting but which do not appear as specific items on the agenda of the present meeting. The chair must ensure that only items which satisfy those conditions are discussed at this point of the meeting. Usually this is a brief item, because if substantive discussion is anticipated then the subject would deserve to be a main agenda item.
5   *First main item:* This should be given a descriptive title, of course. Give details of supporting papers, or say that a presentation will be given at the meeting. Indicate who has the lead responsibility for the item. Other main items follow.
9   *Any other business:* In theory, this allows members of the group to raise topics for which there was no place earlier in the agenda. By using item 4 properly, one category is removed. Other possibilities are subjects not thought about previously, or which have arisen during the meeting's discussion of the main

items. In practice, people are getting ready to go at this stage. There is not much sympathy for someone who introduces a complex topic which would have been better dealt with earlier. However, this item is a necessary safety valve.

10 *Date of next meeting:* For a committee which meets regularly it is important to fix this. Some committees plan their meetings a long way ahead to prevent clashes. If a date is not chosen it can be very tedious to negotiate afterwards a date acceptable to all concerned. This unhappy job usually falls to the secretary.

Within this formal framework you have decide the order and timing of the main agenda items (5–8 in the example). Sometimes there is an obvious order suggested by the shape of the meeting (as Figure 6.3). In other cases the items are less interdependent. In that case, consider the following:

- Is it better to have long, complex items before short, easy ones? If the meeting starts after morning coffee and the chair expects that the members will know each other and be on top form, perhaps this order would be best. In other circumstances, it can be best to warm up on a relatively easy item before tackling the difficult ones.

- Try not to deal with contentious items last. Even if you expect what is politely called a robust meeting, then it can help to have a few easy or noncontentious items so that everyone ends feeling more benevolent. This is important if there is a social event afterwards.

It is difficult to estimate timing, but even approximate times are helpful for the reasons given earlier. Experience is the best guide, coupled with a judgement about which items are likely to be contentious.

The venue of the meeting can be important. It includes both the geographical location and the particular room where the meeting is to be held. Take location first. For some regular committees there is one habitual venue. Nowadays, people are more inclined to question this. There is often good reason to hold meetings near the operational workplace which may be affected by the decisions of the committee. It is also a good idea to share the burden of travel, and be seen to move about between company premises. For occasional (one-off) meetings, or meetings with contractors (whether suppliers or customers) the same factors apply as for presentations.

The room chosen for the meeting must be able to seat the necessary number of people comfortably. If a presentation is included, there needs to be easy transition between concentrating on the focus of the presentation (a screen or a lectern, perhaps) and moving into the meeting mode. These requirements can conflict. For the presentation people have to face approximately the same way; for a meeting they have to face each other. It may therefore be necessary to have a short break while the furniture is rearranged. In that case try to minimise the disruption.

Besides the timing of each item within the agenda, you need to decide the start time and end time for the meeting. There are some practical constraints and some psychological factors to consider. The practical constraints include travel arrangements for participants, whether the meeting is associated with a site visit or a social occasion like a meal, and the availability of the venue. The psychological factors include the fact that most people perform best in the second half of the morning, and that people will wish to get away on time at the end. The last mentioned can be a useful help to conducting business briskly, but make sure that it does not prevent you doing it thoroughly.

With the agenda settled, the chair needs to plan the conduct of the meeting. There are several jobs to be allocated:

The chair (or co-ordinator): Consider who should take this role. It need not be the most senior person present, or the person on whose initiative the meeting has been convened. Sometimes that person prefers to sit aside, unobtrusively, and delegate the running of the meeting to somone else. If they do that, then in fairness to the chair they must not intrude in the running of the meeting except in a dire crisis.

Secretary: This role is essential, except in the very simplest and smallest meeting. The secretary deals with the formalities already mentioned, and also keeps the notes from which the record of the meeting can subsequently be written. Sometimes this has to be done and agreed during the meeting; this is a considerable pressure which should not be given to the chair.

Scribe: It may be necessary to use a flip-chart or OHP during the meeting, to record ideas as they arise in discussion. This is especially important during brainstorming, examined in Section 6.4. In such cases the person who writes on the flip chart or OHP on behalf of the group is termed the scribe. Standing up and using the pen is quite a powerful and demanding role. Neither the chair nor the secretary of the meeting should usually do it. It would detract from their ability to carry out their primary function. Moreover, it would make their role too prominent, so that they dominate proceedings in a way which members might resent.

Timekeeper: In every meeting there should be someone who keeps a careful eye on time, in relation to the planned timing for the agenda. Often the secretary does this. In a complex meeting it is good to give this task to someone else. They should remind the meeting as a whole, or the chair, when target times have been reached or passed.

These roles should be filled by people who can work together as a team. Indeed the whole meeting should be an example of teamwork. In planning it you should therefore consider the way management teams work, and the roles which are needed. If you are not familiar with this subject, then you should look at the work of Meredith Belbin (1984 and 1993). He identifies nine roles needed in a management team:

1   The chair (or co-ordinator), who can control and organise proceedings. This requires the ability to keep a 'helicopter view', and diplomacy in managing the conduct of the meeting.

2   The shaper, who has a clear overview of the proceedings and feels the need to influence their shape. The chair needs these skills, of course, but other shapers can come into conflict with the chair and with each other.

3   The plant, who is good at providing bright ideas and fresh insights. Plants tend to be bored when it comes to discussing the practical consequences or implementation of the ideas.

4   The monitor–evaluator (or critic), who is good at judging the value and feasibility of ideas and plans of action from a practical point of view.

5   The company worker, who can help achieve a practical and workable plan arising from its decisions.

6   The team worker, who can help the group work more effectively together.

7   The resource investigator, who will identify external resources and contacts which may help the group achieve its aims.

8   The completer–finisher, who will make sure that the group completes its agenda with the actions arising from each item clearly defined and agreed.

9   The specialist, who provides expert or technical knowledge on an individual subject which the group is addressing.

Most people have some professional or other specialist knowledge in particular areas. Nobody has it in all areas. Thus we should act as a specialist (role 9) only within our relevant field of expertise. The other eight roles are more general. Each individual will have natural skills for one or more of them. These skills can be measured by applying psychometric tests, but that is rarely available to us for particular meetings. More practically, the chair should assess the effective roles of those present. This will help the chair to allocate people to the jobs mentioned, and also judge whether conflict or difficulty is likely during the running of the meeting. Here are four types of problem:

- Having a person in the chair who is not suited to that role. This is exacerbated if there are others in the group who could perform the role better.
- Shapers who wish they were in the chair can compete negatively with the actual chair of the meeting.
- Plants who vie with each other when producing ideas. They can get very competitive about this.
- Team workers who get frustrated by what they see as an undue task orientation by the rest of the group.

The chair should consider the people attending, and decide who should best undertake the various jobs for the group. After the chair, the secretary is the most important. You must decide what the secretary should produce as the formal record of the proceedings of the meeting. It might be

1   A simple list of follow-up actions agreed. This is suitable for simple, regular progress meetings about particular projects or departments.

2   More formal minutes which have the same headings as the agenda, and record decisions and actions. In the case of actions it shows who is responsible for carrying them out, and by when. This is the preferred style for most

minutes. There is no detailed record of who said what, or how decisions were arrived at.

3   As for no. 2, but with some information about the discussion which took place. This should be recorded in the past tense, as recorded speech. If there is fundamental disagreement on an important topic, people sometimes require that their minority view be recorded. Always respect that wish. Only use this style of minutes for important meetings, where the rationale for decisions may need to be reviewed later by people who were not at the meeting. Avoid it for normal business.

The output of the meeting is a record in one of these three styles. The secretary must know which is required. Similarly, the chair must brief other people who will undertake specific jobs or team roles during the meeting. It might even be worth having a chair's premeeting, to discuss the handling of the real meeting. Be careful, not to spend too much time talking about the detail of the agenda items; that leads to endless meetings about meetings! The purpose is really to plan the 'process' aspects of the meeting, rather than the 'task' aspects. The distinction between task (the words) and process (the music) was drawn in Section 3.2 and Figure 3.3.

The final aspect of the meeting to be planned is the practical one of logistics. For a big meeting you need to liaise with the administrative staff responsible for the building and room concerned. Think about refreshments. Should these be provided, and if so before, during or after the meeting? That is partly determined by the facilities available. One thing is certain: the arrival of tea or coffee during a meeting is a major disruption. Serving beverages and passing round biscuits diverts attention. It is particularly annoying for whoever is leading the discussion at the time. It is better to pause at a natural break in the meeting to take refreshments, or at least to serve them out and bring them back to the meeting table. It is best to decide the policy on this beforehand, but an ad hoc decision from the chair may be needed on the day. The other aspects of logistics are considered in the context of presentations, in Section 5.3. If you want further information, see Janner (1986) and Seekings (1989).

6.4 Running a meeting

General Eisenhower once remarked that plans are nothing, but planning is everything. That applies to business communication as well as to military operations. Your plan for the meeting should not be rigid; you must expect to have to adapt it on the day. If everything were clearly determined then there would be no need to hold the meeting at all! However, by doing detailed planning you will have a better foundation for being flexible on the day without losing control of the overall proceedings and purpose.

For a large meeting people will have been allocated their jobs as chair, secretary, scribe and timekeeper. Together they form a team. For small meetings the team will probably be just the chair and secretary. But in either event this team should assemble early and check the administrative details well before the start time of the meeting. It is a good idea for them to put their own papers at their respective places in the meeting room. This will stop other people accidentally taking a reserved place. Only occasionally is it necessary to have named places for everyone. It is always useful to have the secretary beside the chair.

The team should receive the members as they arrive. If refreshments are being served this is a useful opportunity to make introductions, break the ice, and check details of procedure with key people attending. Anyone who has to mount a presentation within the meeting will wish to follow the guidance given in Chapter 5. So make sure they have good opportunity to check visual aids, microphones and other facilities.

For a large meeting, the team will need to work together. Sometimes, with important visitors, you might ask individuals to 'host' specific guests. Normally, however, things are less formal. However, one of the problems is then remembering people's names, especially if you are introduced to several people in quick succession and have, in turn, to introduce them to other people. There are two useful tips here:

— Have in your pocket a small piece of paper, folded to no larger than A7 size. This is 74 × 105 mm, or an A4 sheet folded in half three times. This is small enough to be retrieved discreetly from your pocket and read in the palm of your hand. On it write the name of important people you need to look out for, or newcomers you expect to meet. As with most notes, the act of writing this out will help you to remember the detail. You may not need to use the note, but it is reassuring to have it there in case. Sometimes, in the pressure of the moment, you can forget even familiar names.

— Remembering names is a problem for most of us. Sometimes you can get photographs beforehand, but that is exceptional. Usually you have to use one of the standard techniques. Repeat the name when you are introduced to a person, to make sure you heard it correctly and to help imprint it in your memory. Try to remember the two main initials, to form a memory bridge between the face and the full name. Use the name in subsequent conversation with the person. If the nightmare happens and you need to introduce someone whose name you have forgotten, then be honest and make sure that the joke is at your own expense. People are nearly always sympathetic, because this problem happens to us all sometimes. The problem gets worse as you get older; this can be the basis for a self-deprecatory excuse.

Eventually the appointed hour arrives, and you must call the meeting to order. Someone with a loud voice, or access to the public-address system, is useful for that. Get people seated quickly and allow them time to get out their papers before starting. Never delay the start by a more than a few minutes, even if someone important is missing. It is better to start on time and rearrange the order of the agenda than keep everyone else waiting around wishing they could be doing something useful.

After bringing the meeting to order the chair will usually extend a welcome and perhaps make some pleasantry to relax the atmosphere. Occasionally, it may help to remind people of the main reason for the meeting and instil a sense of commitment or even gravity to the proceedings. Either way, you are now embarked upon item 1 in the sample agenda considered at the start of Section 6.3. From this point on you reap the benefit of your careful planning of the agenda.

Some meetings are purely mechanical and formal. These are largely a waste of time. Useful meetings involve discussion about new ideas, leading to new conclusions or decisions to which the group feel commitment. These are the really useful parts of a meeting and they are often buried within the overall structure which we exemplified in Figures 6.2 and 6.3. It is the chair's job to lead the meeting through these discussions without unduly influencing the outcome. For each of these critical phases, the shape can be represented as in Figure 6.4.

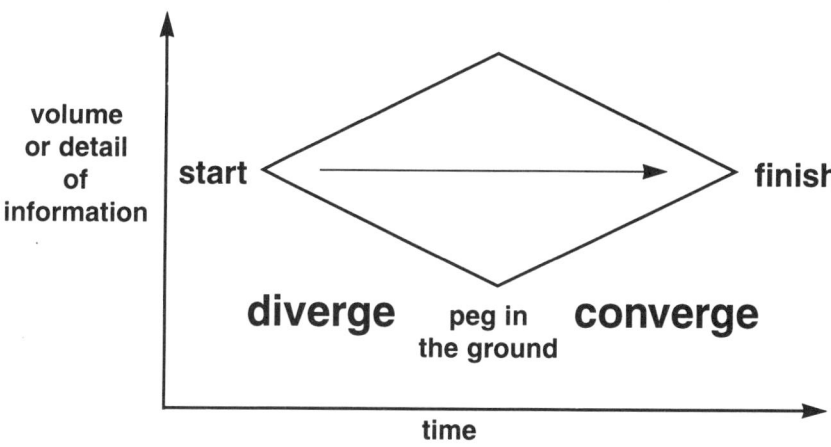

Figure 6.4   Diverge–converge process

To have a fruitful discussion the group must agree where they are starting from and what they are trying to achieve. That sounds obvious, but you cannot assume that everyone in the group shares the same perspective. At the start of the discussion phase the chair should therefore state the start point and aim explicitly. This might involve summing up a presentation (or other input of information), defining the problem it reveals and stating what the meeting should aim to achieve in its impending discussion. Such a preliminary statement of the start point helps everyone to focus their mind. It gives opportunity for any misundertanding to be resolved without wasting time in fruitless discussion. Groups often omit this step, in their haste to get to grips with the problem. The chair should make sure it is not omitted.

The next stage is to broaden the discussion. This is the diverge phase. It involves receiving fresh ideas, and perhaps supplementary information or knowledge relevant to the problem. The group should listen to these ideas and try to build on them, so that the argument is carried forward and achieves a better result than any one member could achieve on their own. That is the measure of success of a meeting: the synergy which makes group performance exceed any individual performance. It is not easy to achieve, especially amongst people who do not know each other and work regularly together. But it can be done, and it is a vital function of the chair to make sure that it is done.

One of the fastest and most effective ways to broaden a discussion is brainstorming. This supplies a stock of fresh ideas which the group can then appraise and build upon in its discussion. However, the process of brainstorming has to be managed carefully if it is to be fruitful. It must also be used with caution:

- Don't get a group of nonexperts to brainstorm a technical problem. You need a specialist with the relevant expertise. This is where an old adage applies: if you want to jump over a two-metre fence, you need one person who can jump two metres, not two people who can jump one metre.
- Don't use brainstorming unless every member of the group understands the rules and undertakes to comply with them. This means, for example, setting aside hierarchical relationships which apply outside the session.

Thus if you decide to use brainstorming it must be by mutual consent. The chair, on behalf of the group, must then ensure that the rules of brainstorming are upheld. Here they are:

1   Make sure the group is comfortable and is mentally warmed up. It is no good trying to brainstorm when people have just met each other, or if everyone is desperate for lunch.
2   Have a scribe ready to record key ideas on a flip-chart or whiteboard. Make sure there is plenty of room for this. In the case of a flip chart, it should be possible to tear off and display sheets as they are filled up.
3   Participants must think freely and throw in ideas without inhibition. The ideas should be encapsulated or suggested by a single word or very short phrase.
4   The scribe records the ideas in the form given. At this stage, nobody is allowed to comment on or criticise any of the ideas.
5   Participants should feel free, and should be encouraged, to build on earlier ideas. Thus a new idea can embrace an earlier one, or go off at a tangent. The ideas do not belong to the individuals who offered them; they become the property of the group as a whole.
6   This phase of free idea generation should proceed for about five to ten minutes in most situations. The chair should judge when the flow of ideas is drying up, and announce the end of that phase.
7   The chair now leads the group into a more critical phase where it will appraise the ideas which have been generated. This is best done by grouping related ideas together. The scribe can do this, under the direction of the group, by putting coloured marks or circles round the related ideas. Typically four or five such groupings will emerge. The group then looks at each related set in turn, discussing the ideas and trying to distil the best of them. There may be some development of the ideas to express them better or make them more practical.

8   The distilled ideas are then listed on a fresh sheet or whiteboard. They are the tangible output of the brainstorming session.

If carefully managed, brainstorming can produce a wealth of fresh, relevant ideas in a very short time. Those who have not experienced it before are often amazed by the results. The process itself follows a diverge–converge sequence, but overall it contributes to the diverge phase of the discussion as a whole.

The chair can then encourage the group to continue the diverge phase of their discussion. This may involve refining the ideas from brainstorming, introducing the hard practical constraints and working towards practical options for solving the problem. In the team-role terminology of Belbin (1984 and 1993), the plants had a field day during brainstorming; it is now the turn of shapers, monitor-evaluators and resource-investigators (in particular) to make their distinctive contributions.

The diverge phase of the discussion must not be allowed to continue indefinitely. That might be interesting but the meeting would never reach any agreement. At some point the meeting must consciously shift to the converge mode. Here, it will appraise ideas more critically, reject some in favour of others and gradually work towards an agreed decision. Again, the chair has to judge when the right moment has come to change from diverge to converge. The chair should sum up where the discussion has reached, what options are still under consideration, what has yet to be achieved and that the whole approach now has to shift. This is the 'peg in the ground' marked on Figure 6.4. Thereafter, the chair must be firm that old issues are not reopened; the group has to converge in a controlled way to reach a decision.

There are several ways for a group to make its eventual decision:

1   By the boss listening to what is said, and then giving a ruling. The boss, of course, may or may not be in the chair.
2   By taking a vote at the end of the debate, with everyone having agreed to adhere to the outcome.
3   By seeking consensus, where all members accept and feel committed to the group decision even though some of their personal preferences may not have been adopted in that decision.
4   By seeking unanimity, where all members are in total agreement about the group decision.

Think about meetings you have attended where groups have arrived at decisions. Which of the four methods was used? Which is most difficult and time consuming? Which results in the greatest commitment of the members to the resulting decision?

Think carefully about those methods and the questions. There may be a legitimate place for each method, in particular circumstances. But here is some general (and therefore highly dangerous) comment:

● Voting is divisive. People resent being over-ruled. The victors are jubilant; the vanquished are resentful. The group as a whole does not 'own' and feel committed to the decision. A substantial minority may even hope, secretly, that it will fail. If it does, they will be ready to say 'Well, I never agreed with it; I voted against'.

- A decision by the boss, unless very persuasively argued, can be equally divisive. Occasionally, such a decision is needed. But in that case, why have the meeting and the discussion? If the boss's view is so logical, why can't it be used to reach a consensus?

- Voting and decisions by the boss are quick and clear. However, in most situations that does not outweigh the disadvantages.

- Unanimity is very difficult to achieve, and often impossible. Save in exceptional circumstances (perhaps involving life-or-death decisions), it is unrealistic to aim for unanimity. The group should recognise and accept this.

- In practice, it is hard enough to reach consensus, but it is a goal worth aiming for. It does result in ownership and commitment by the group. And it can be achieved in a realistic time, given good group behaviour and a sensitive chair.

Normally, then, the chair will be aiming to help the group reach a consensus decision from its discussion. That is what the 'Finish' means at the end of the converge process in Figure 6.4. It involves people shifting from their predetermined views, or prejudices. It requires people to give way in the face of facts, logic or persuasion. It requires people to present their case logically when they can, and persuasively when they can't; remember the motto from Section 3.4. During the converge phase, Belbin's (1984 and 1993) company workers and completer-finishers have a special role to play. It is sometimes remarked that Japanese business meetings seem interminable to westerners because the Japanese are seeking consensus, and commitment in the fullest sense. Sometimes western business meetings are too task oriented, and want to make a quick decision so that the action can start. As a result the decision may have only the appearance of consensus. It is hardly surprising if the divisions become apparent later, during the implementation phase.

All this has an effect on the chair's management of the converge phase of the discussion, not only in what is done but in how long it takes. In the (hopefully) rare cases of boss's decision and voting, the converge phase is quick and easy. It is scarcely surprising that the result is less valuable! In the (hopefully) equally rare case of seeking full unanimity, a lot of time must be allowed: far more than for the diverge phase of the discussion. However, in what should be the modern norm of team working leading to a consensus decision, the chair should allow at least as much time for the converge phase as for the diverge phase. If there is a time limit on the discussion (as there will be in a well planned meeting) it is wise to allow 40% of the time to diverge and 60% to converge. Remember that to diverge is quick, easy and fun; to converge is slow, difficult and painful. Don't run the risk of running out of time before the group has reached its all-important consensus.

The chair must manage these key periods of discussion very carefully. They are the value-adding nuggets of the whole meeting. It helps if the chair, and the group as a whole, understand the diverge–converge model. It is simple enough to explain, even within an agenda item, and provides a shorthand for the chair to control proceedings. The diverge–converge process starts with a simple aim and ends with a simple decision. In between it can be messy and difficult. But it is worth it. As Oliver Wendell Holmes (1841–1935) remarked 'I would not give a fig for the simplicity on this side of complexity, but I would give my life for the simplicity on the other side of complexity'.

There is another problem which the chair is sometimes confronted with. This is when the group does not 'spark' on the topic. The chair introduces the item, outlines the start point, but the group does not get going on the diverge phase. There are two possibilities:

— The group genuinely has no ideas or relevant expertise to bring to bear. This situation should have been identified during the planning of the meeting. Perhaps a different set of attendees was needed: more senior, or with special expertise. If this situation arises, the topic may need to be referred to a different group or addressed outside the meeting.

— The group has not warmed up, or people are reluctant to make the first move. Often, it only takes a couple of contributions to get things moving. The chair can do two things. One is to keep silent. In our frenetic business life, silence is rare. It is therefore powerful. Allow it to continue for two minutes. Usually someone will say something and release the logjam. If they don't, the second step the chair can take is to make a contribution themselves. Normally this is discouraged; the chair should assume a neutrality which allows them to retain the respect of the group during what can be heated debate. In this situation, an idea from the chair can help. Of course, it need not reflect the chair's own personal viewpoint; it can be offered as a maverick idea to get discussion going. That can be done without weakening the chair's standing within the group.

To cope with all these eventualities it is good if the chair can think about each of the discussion phases of the meeting and identify a possible outcome. That possible outcome need not reflect the chair's own preference (although it might!); but it should be a defensible view which can be offered to the group in the unhappy event of the group failing to identify and refine its own. It's all part of being a good chairperson.

The secretary, meanwhile, has been taking notes of the proceedings. That is why a different person is needed for each job: chair, secretary, scribe and perhaps timekeeper. As we discussed in Section 6.3, it will have been decided which type of record of the meeting is required. The secretary will be able to keep notes to an appropriate level of detail.

Occasionally, it is necessary to record an exact form of words verbatim. This might be a formal resolution or a carefully worded decision. The secretary should, through the chair, read such wording back to the group to ensure that the record will be correct. Less often, a meeting will insist on the whole of the action sheet being agreed in this way before they close the meeting.

In any event, the secretary should aim to agree the formal record with the chair within two working days. The record should usually then be sent to all those who attended the meeting, plus those who were invited but were unable to attend. Sometimes information copies go to other people as well. Timely issue of the record is an important part of ensuring the overall effectiveness of what the meeting achieves.

Planning and running a meeting is a science and an art. As with other aspects of management, success comes from thoughtful application of principles as you build up your experience. Throughout I have stressed the importance of those attending the meeting working together well as a group or team. That is a major factor in deciding who should attend. It is also something which can be nurtured in groups which meet together regularly.

It takes time and effort to develop people into effective working groups or teams. It is not appropriate in this book to look in detail at the process, beyond recording that there are usually four stages:

1   *Forming:* This is the recently formed group who are politely getting to know each other and each other's capabilities. It is the undeveloped team.
2   *Storming:* This is the group which is trying out different ways of working, to find which is right for them and the problems they face. This is the experimenting team.
3   *Norming:* This is the group which has found the answers to some of those questions and is establishing accepted methods of working. This is the consolidating team.
4   *Performing:* This is the group whose members know each others strengths and weaknesses well, are confident and committed to group success and enjoy a sense of belonging. This is the mature team.

Figure 6.5 shows some of the symptoms of group behaviour which are associated with each of the four phases.

Some of the signs of development which are visible to the chair and other members of the group are

* from polite formality, to conflict, to the controlled resolution of conflict
* from speaking to listening to thinking
* from task orientation to balanced management of task and process
* from having little commitment to a full commitment to team success

The chair, as a manager, will try to recognise the symptoms and nurture the group towards a higher level of achievement. Across the top of Figure 6.5 is a scale from 0 to 12. This is an unrolled clockface. It provides a shorthand for the group to identify where it has got to. For example '6-o'clock' is the start of the norming phase. Every group has relapses as well as periods of progress. But regular meetings will help the group to develop. Meetings of mature teams are efficient, effective and even enjoyable. Have you ever experienced this? It's a goal worth seeking! Remember the acronyms for TEAM: together everyone achieves more; and together excellent, alone mediocre.

6.5 Summary

This chapter has considered meetings which are convened deliberately, away from the operational business environment. In general, such meetings have four tasks:

	0	3	6	9	12
	Forming the undeveloped team	**Storming** the experimenting team	**Norming** the consolidating team	**Performing** the mature team	
	* tense; polite	* opening up; conflict	* confidence	* commitment	
	* very little listening	* listening and thinking	* develop ideas	* relaxed development	
	* stick to brief	* agree procedures	* flexible approach	* refine group procedures	
	* confusion about task	* rigid view of task	* see task in context	* manage task and process	
	* decisions by leader	* majority views	* seeks consensus	* achieves consensus	
	* conceal weaknesses	* more self awareness	* some mutual support	* support for members	
	* group is defensive	* group is insular	* group more outgoing	* group is responsive	

Figure 6.5   Team development

to receive information; think carefully about it in discussion; reach decisions; and then plan the implementation of those decisions. The scope of individual meetings will vary, but these four tasks together provide a general template.

The overall shape of those four tasks is converge–diverge. However, when each item is dealt with in the meeting the chair needs to introduce it, state the aim to be achieved, and then give a summary at the end. For a complex meeting with several major agenda items, the chair must do this for each such item and also for the meeting as a whole. This is an important and challenging responsibility.

The plan thus far will enable you to decide who should attend, and prepare an agenda to send out beforehand. You must also decide how the starting-point information is to be provided for each item. You may have to ask someone to assemble it and write a briefing paper to circulate with the agenda. You may need a presentation at the meeting itself. You may decide that the whole item should be based on a wider management report or paper. In that case be careful that the role of the meeting has not degenerated into rubber stamping. Supporting papers should follow the principles given in Chapter 4. This means that even if the meeting does not accept the entire paper it can still build usefully on it. The original author, or someone else, may be asked to amend and develop it in the light of fresh guidance.

With the purpose, aim, scope and structure (PASS) of the meeting decided, the chair and secretary can make the detailed arrangements. The first task is to complete the agenda. In the case of a committee or other formally constituted body this should follow the established format. In any case the main items should be arranged with care, and approximate timings shown.

The choice of venue (location and room) will involve similar factors as for business presentations. However, for a meeting people need to face each other easily. If a presentation is included within the meeting then this can involve a compromise. If this is difficult, adopt the seating arrangement which favours the meeting.

The time of day for the meeting will often be chosen for practical reasons. However, there some tactical considerations such as having a backstop which ensures the meeting will not go on too long. It helps to show the estimated end time for the meeting clearly on the agenda.

There are several important jobs to be carried out during the meeting: chair, secretary, scribe and timekeeper. In a complex meeting they should be done by four different people. They, and the other participants, have to work together as a team. It is worth considering Belbin's team roles, and the strengths of each individual to carry them out. The chair should allocate the jobs and consider what to do about any shortcomings in the group, or possible sources of conflict.

It is important to decide which of three styles of record is appropriate: list of actions; minutes which show only decisions and agreed actions; or fuller minutes which record the main flow of debate and thus the reasons for the decisions made. The secretary needs to know which is required. People with other roles need to be told. The chair may decide to hold a premeeting to discuss these 'process' aspects of the meeting.

The final aspect of planning is the logistics. This includes the practical arrangements and the serving of refreshments. It is important to ensure the last will not disrupt proceedings.

Even with a good plan, the running of a meeting is not easy. The main responsibility rests with the chair, supported by a team of at least the secretary and perhaps the scribe and timekeeper. On the day, they must ensure that the preparations are complete and receive the people attending. In a sense they act as hosts, with the attendant hazards of protocol and remembering names. Start the proceedings on time. Work to the plan of the agenda, which all those present should understand and be familiar with.

Identify the important periods of discussion which add real value to the meeting. Manage these according to the diverge–converge model. Consider brainstorming to achieve a stock of fresh ideas, but if you adopt it make sure the nine rules are adhered to. Normally, in the converge phase you should be aiming for consensus decisions which have the full commitment of those present. This will require at least as much time for the converge phase as for the diverge phase. To run out of time undermines the whole achievement. Make sure that the record reflects what has been done, if necessary agreeing the secretary's words in the whole group. Make sure that the record is issued within a few working days to those with a direct interest, whether or not they were present at the meeting.

Finally, in a group which meets regularly, think about their team development. There are four stages: forming, storming, norming and performing. Learn to recognise the symptoms of each, and encourage the group to form its own view of its progress, in terms of the clock face. By focusing on group process as well as the task, you will get much more effective results from them. The experience could even become a worthy reflection of your management skills!

Chapter 7

Using information technology

'The telephone may be appropriate to our American cousins,
but it is of no value here, since we have an
adequate supply of messenger boys'
The Times, *London, 1900*

7.1 Introduction

At this point it is worth recalling the overall structure of this book, as discussed in Section 1.3 and illustrated in Figure 1.2. Recall that Chapter 2 looked at the communication process, the business enterprise and what is meant by management. Those ideas were the foundation for discussion of individual types of business communication in Chapters 3–6. That discussion is now complete. This chapter has a quite different purpose.

So far the various types of business communication within the context of a management problem have been considered. The purpose, aim, scope and structure (PASS) of a business communication were defined by reference to the management learning cycle shown in Figure 2.8.

Now take a slightly different view of a business communication. Consider it as one transaction within the total 'business information system'. This will help understand the increasingly important role of information technology in support of that system. This section gives a brief overview of what is meant by the business information system, and the emerging role of IT to support it. The following sections then examine how we can use IT as a personal aid, how we can use it within the wider business enterprise, and what the trends are for the future. If you are familiar with modern IT developments you may wish to go straight to Section 7.2 now.

The term 'information system' means different things to different people. To an IT professional it might mean computer hardware, software and communications. This is the narrow view of the term. As a manager you need to take a wider view. The 'business information system' which you need to consider includes people and business procedures, as well as the technology. Remember: people, procedures and technology. It must be designed to collect, store, process and communicate information, so that people can make decisions and initiate action. That accords with our earlier idea of an information value chain supporting the management learning cycle. The business information system must be able to handle both internal and external flows, to support both the organisation and its operational task:

— Internal information flows: Different types of information may flow top-down or bottom-up within the organisation structure. There are often formal procedures for this, such as reporting procedures (upwards) and cascade communications (downwards). There will also be horizontal flows of information, directly concerned with achieving the operational task. Think about this, in the context of your own organisation. You may have been involved in business process re-engineering (BPR). This is a technique which tries to adapt business processes, and the related flow of information, to match the operational task more closely.

— External information flows: The business enterprise does not work in isolation. It exists in an environment which includes customers, suppliers, competitors, regulatory authorities, and other 'stakeholders'. Again, think of examples of this, from your own experience. What type of information is exchanged with the various parts of the business environment? Particularly at senior levels, external information tends to be less specific, less formal, and less immediately relevant. However, managers need to absorb it, and use it alongside internal information to support their planning and decision making. Increasingly, information about the business environment and its trends is available electronically, for example from online databases. At operational level, detailed information may be exchanged with suppliers and customers. For example, this is necessary to implement just-in-time procedures which reduce the capital cost of stock. Such links are usually automated, using electronic data interchange. This has removed the need for much direct human, or paper-based, communication.

You need a clear view of the business information system before considering how to use IT. Surveys have shown that there is a low correlation between the level of investment in IT and business success. This is because if you automate a bad business process you will be worse off. If you automate a good business process you may be better off. If you can identify improved business processes, made possible by IT, you may get spectacular benefits. The same principle applies whether you are considering IT as an aid for you personally or for your business as a whole. IT is an enabler for business change, not just something which automates what you happen to do already.

So, what is information technology and what does it offer as an enabler of business change? IT has been defined as the convergence of computing and telecommunications made possible by modern microelectronics. Thus there are three underlying technologies:

- microelectronics, to represent information in electronic form
- computers, to store and process the information
- telecommunications, to transfer the information to where it is needed

Consider each of these in turn, and then the legal aspects of using IT systems.

Microelectronics underpins all modern IT. Integrated circuits (ICs), or 'chips', have many transistors and other components fabricated on a small piece of material such as silicon. The capacity of ICs has developed very rapidly, at about 30% per year since 1960. It is now possible to put many millions of components onto a chip, and this rate of development is set to continue. Chips are best able to handle data which are in digital form, using numbers (or bits). Properly designed digital technology will handle data at very high speeds and virtually error-free. This feature has made complex computers and high-quality communications possible. The convergence of these technologies gives modern IT its unique power.

Computers manipulate data (numbers) in accordance with a set of instructions. The result needs to be presented in a form which the user can interpret as meaningful information. The physical machine (hardware) is controlled by a set of instructions (software). The operating system tells the machine how to work, while an application program tells the system to do a specific type of job for a user. Computer hardware has developed in step with electronics: the power/cost ratio has risen by a factor of ten every 6–7 years. Computers have become both smaller and more powerful. After mainframes and minicomputers, the personal computer (PC) had a major impact on business from the early 1980s. The four main applications of the PC have been: word processing; spreadsheets (processing arrays of numbers); storing information in databases; and presenting data graphically. Software packages are available for each of these functions, and also integrated software which allows the functions to be combined, with the data shared.

More recent trends include:

- Graphical user interfaces, with symbols on the screen which resemble familiar items in the paper-based office environment. The user selects the symbols using a mouse or other pointing device.
- Windows environments, where several applications are displayed on the screen at the same time, can be accessed quickly, and can share data.
- Local area networks, which connect computers together so that corporate data can be stored and exchanged around a site in electronic form. 'Client-server technology' is one way of making such services available to users throughout an organisation.

- Laptop or notebook computers, for use by staff on the move. They can use modern communications to link into fixed office systems.
- 'Groupware', specifically designed to support people working together in teams rather than individually.
- Voice-based interfaces with computers. This technology is now set to make a major impact on the way we use systems.

The third element of IT, telecommunications, conveys data or information from one place to another. This involves

- transmission, between defined points in a network
- switching, to get messages across the network to the required destination
- storage/processing of messages for the convenience of users

There are two important trends in transmission technology. The first is the use of glass fibre-optic cable, which can carry many thousands of telephone channels (typically 35,000). FOC is now the main method of transmission between fixed points, over land and under the sea. This has made transmission much cheaper and released the radio spectrum for what it is uniquely good at: mobile communication. Land-based transmitters can reach people and vehicles within a limited area, while satellites have a greater geographical coverage.

This transmission capacity makes possible switched wide area networks, which serve many users over a large geographical area. The public switched telephone network (PSTN) is much the largest. On business premises private automatic branch exchanges extend the PSTN to all users on a particular site. Cellular radio systems extend it to mobile users; such systems have expanded very rapidly since the mid 1980s. The early PSTN and cellular radio systems used 'analogue' technology. This is convenient for transmitting speech but is not directly suitable for digital data. In developed countries most transmission and switching is now digital, but the periphery of the network (connecting users to their local exchange) is still analogue. Devices called modems can be used to send digital data (including facsimile) over an ordinary telephone connection, but speed and quality are limited.

To handle the increasing amounts of computer and other data, users need a wholly digital service. Separate packet-switched networks have provided such a service since about 1980, but the business communication system of the future is the integrated services digital network (ISDN). It is all-digital and its basic channel speed (64,000 bit/s) is at least three times faster than even the best modem connections. ISDN is now available internationally, and supports text, data, voice and image transmission. For example, videoconferencing is much cheaper on ISDN.

Wide area networks convey the users' data or information without altering them. However, it is possible to add computer processing and storage to the network, and offer value-added services. These may provide facilities for storing messages, manipulating them, and sending them to selected addressees. They include the security or password arrangements to allow only legitimate access to the system, and arrangements for monitoring usage and billing subscribers. Examples are viewdata (linking a telephone line with a TV or PC, to make an interactive terminal), electronic mail (e-mail), and multiway text/videoconferences. Internet is a significant development; see Section 7.4.

Finally, third-party information providers can use the system to sell information about financial markets, up-to-the-minute news, and other subjects for which the value of the information depends on having it quickly. In France, the Minitel system has over six million users, and offers services like home shopping and banking from 24,000 different providers.

These growing networks of computers are very powerful. They bring with them certain legal problems. As a user of IT systems, there are two which you should be aware of: the protection of personal data, and the misuse of computer systems.

There is data-protection legislation in many countries to protect individuals against misuse of data about them which is kept on computer systems. An important provision is the right of the data subject to see a copy of what is held about them, and to require changes if it is inaccurate. In the UK the 1984 Data Protection Act established a data protection registrar, and required companies and individuals to register if they process data about living, identifiable individuals. There is a legal obligation to safeguard personal data, both physically and electronically. For example, any transfers of personal data to third parties must be registered.

There are other types of misuse of computer systems. They may be impaired or disabled by viruses introduced deliberately or by innocent third parties. They may be accessed illicitly by hackers inside or outside the organisation. Some countries have made such activity a criminal offence. For example, the UK Computer Misuse Act 1990 defines three levels of offence, relating to penetrating the system, reading data, and manipulating data. The threat is such that most organisations now need a planned and documented information security policy. It should cover both internal procedures and external safeguards.

This introduction has outlined the concepts of a business information system and looked at some of the more important IT developments. This should equip you for the rest of this chapter. However, if you want more information there are plenty of texts about IT and information management. See for example Silk (1991) and (1995).

7.2 IT as a personal aid

Turning now to the more practical aspects of using IT systems there is a lot of affordable technology available. You need to be a well informed and discriminating user of IT!

Just as your company needs a business information system (people, procedures and technology), so you as an individual need a personal information system to support what you do. You have probably evolved your own methods over many years. But even so it is a good idea to take the occasional fresh, critical look at what you do. Think for a moment about how you personally handle information:

- How do you collect information? Where does it come from? Is it formal or informal? Is it structured or unstructured?
- How do you store information? How much is kept only in your head? What do you keep on paper in a personal filing system? And how much in the company's official system? Remember that you can't remember everything. Arthur Schopenhauer (1788–1860) said 'To expect a man to retain everything that he has ever read is like expecting him to carry about in his body everything that he has ever eaten'.
- How do you process information? Do you have to do detailed calculations or other work on hard data? Or do you have to think about a wide range of information, and reach conclusions based on judgement and experience?
- How do you communicate information to those who need to know and act upon it? How do you make sure that those who need it get it? And that those who don't need it don't get it?
- To what extent do you use IT systems to support these four critical activities? Could you use IT more? Do you know enough about the IT opportunities to form a view on this?

Think about your answers to these questions. It may be that you need to redesign your personal information system. This might (or might not) involve a greater use of IT. There are three situations that can arise:

1 Your job is closely linked with a corporate IT system, which you need to use to do most of your job effectively. Some organisations have got rid of paper altogether for business transactions and routine communication and reporting.
2 There are some IT systems available to you, but the extent to which you use them is largely a matter of choice for you.
3 There is little in the way of IT available to you at work. You must make your own decisions about it. If you can't persuade the firm to pay for what you believe you need then you might even have to pay the bill yourself. You must decide whether this would be a good investment.

In the first case, you need to learn and use the system! The firm, of course, should give you training and ongoing support. In the second and third cases you must weigh things more carefully. Consider these issues:

— Have you got a secretary or other assistant who can use IT sytems? If so, is it worth you acquiring equivalent or complementary skills?
— Could you do the various tasks (notably word processing) yourself? If so, would the use of the keyboard slow you down when creating new documents, or when amending existing documents? Section 2.2 showed that the effective speed of composing a complex document is much less than the speed at which most people can write or can soon acquire on a keyboard. However, you need to write in bursts, so that a greater speed is more beneficial than you might at first think.
— Is is worth investing time and effort to learn to use software packages, and get a good speed on the keyboard? Despite what you may be told by some

enthusiasts, there are no short cuts. Modern IT systems are much more user-friendly than they were, but they are also more complex in what they do. Get advice from other nontechnical users about how long it took them to acquire a basic proficiency, and whether they now feel it was worthwhile.

— Apart from the needs of your immediate job, is it worth developing new IT skills so that you are ready for new developments, can influence them for the good of your firm, or enhance your own career potential?

— Are you aware of the increased flexibility which modern screen-based working can offer? If not, get a computer-literate colleague (not a technician!) to demonstrate this, and recount their own experience.

— Which modes of information handling (text, data, voice or image) do you have to handle most? Is it worth acquiring skills for one mode rather than for the others? The one which most people start with is text, using word processing packages.

— How do you need to communicate the information from and to other people? Are they equipped with compatible IT equipment?

— Do you need to work from more than one location? This might include occasional or regular home-working, working from customers' and other outside locations, or from your firm's other offices. Mobility is possible with modern IT, but has to be planned for.

— The issue of standards is particularly difficult. It can be very expensive for a firm or an individual to maintain compatibility with several IT systems (or even one evolving system).

If you have access to a personal computer, or a network terminal with similar facilities, then you should consider each of the main types of application package:

Word processing: This is the most obvious tool. Most people take to it cautiously. They may start by two-finger typing and generating only the simplest of text formats. Then they may share the task with a secretary, preparing a longhand draft which the secretary types up, but then doing subsequent amendments on-screen themselves. Eventually, they may start to both draft and edit text on-screen themselves. This gives a flexibility which few people abandon once they have experienced it! There is then a real incentive to acquire touch-typing skill, to increase speed.

Spreadsheets: These are very effective for presenting detailed numerical data, doing simple statistical analysis and presenting the results in graphical form. An example was given in Section 4.2.

Graphics: These packages can be used to generate visual aids or to enhance the basic graphical output of a spreadsheet package. Figure 7.1 is an enhanced version of the graph shown in Figure 4.3.

Desk-top publishing: These packages combine the features of word processing and graphics, enabling you to arrange text and diagrams flexibly and add artwork which is held in a software library. DTP can be very time consuming to learn, and the amateur tends to produce a bizarre and over-busy layout

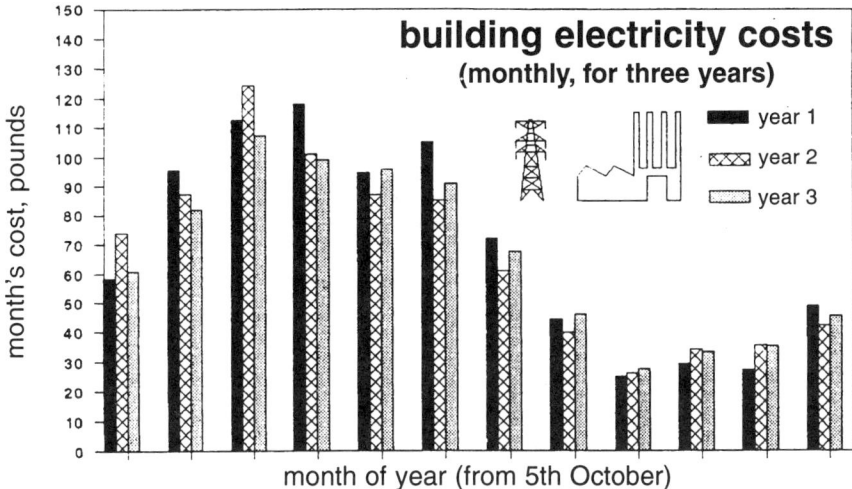

Figure 7.1   Graphics presentation

which can actually detract from the main business message. DTP is best left to the professionals.

Statistical and mathematical packages: These are needed for complex analysis, beyond the capability of a basic spreadsheet program. They are often specialist in nature (for example SPSS for the social sciences). It is rarely necessary for an individual to acquire such packages; they should be provided as a corporate resource where the job makes this necessary.

Computer hardware and software are each developing very fast. There are major competitive battles to establish industry standards both for operating systems (for example Windows and Unix) and for application packages. Related to that is the emergence of integrated suites of software where a single vendor provides all the functionality required in the normal office. Beware of making personal investment in industry or company standards which are liable to change. Keeping up can be expensive and painful!

Since most people start with word processing, some further comment in that area may be useful:

— Learning to touch-type is a significant investment. If you decide to, then the sooner you do so the bigger the reward you will reap. The standard Qwerty keyboard layout was designed more than a century ago, to slow people down and prevent the mechanism of mechanical typewriters from jamming. Almost any layout of keys would be ergonomically better than the one we actually use! Nevertheless, the Qwerty standard is established and we have to live with it. Even with the greater use of voice input to computers, it is likely that the keyboard will still be needed to prepare more complex documents. There are many training courses available; most people can achieve 30 words per minute after only 30 hours of tuition. There are computer-based typing tutors which help you develop and maintain your speed.

— Spell checkers are a useful feature of word processing packages. They count words and indicate dubious spelling, either on completion of a document or while you are typing. In most cases they suggest a likely correct spelling which you can select quickly. Make sure that the spell-checker's dictionary is in appropriate language, such as non-American English if you are working in the UK. Also ensure that you can add your own terms to the dictionary; this is important when you use technical or professional terms regularly.

— More recent packages offer grammar checkers which are supposed to monitor your style of writing and provide advice. For example they might warn you if nearly all your sentences are long, or if you use the passive voice excessively. Some people find them helpful. However, it is probably better to learn to write well without them. You certainly should not rely on them, as at present they can check only a very few aspects of what we call style. Some present an ease-of-reading index. Grammar checkers can make you think, but they cannot think for you.

— In our international business environment, it can be important to be able to use exotic symbols and accents. Most modern systems and printers support these. There is a view that international business English (in particular) will contribute to the death of many traditional languages. There are about 6,000 languages spoken at present, and it is thought that between 20 and 50% of these will become extinct during the next century. There is also a view that as IT systems become multimedia, we shall be using images much more than text in the future. However, the business use of text is certainly not yet in its terminal decline!

Finally, some advice from personal experience. I have used PCs quite intensively since 1984. I have learnt the importance of keeping backup copies of all the work I do on the PC. Failure of a PC, particularly of the internal hard disk, can be a catastrophe. I therefore keep all my current work on diskettes (floppy disks). I make regular backup copies from the diskettes to the hard disk of the PC (or to the network system); this takes only a few minutes, and should be done regularly. Using diskettes makes things slightly slower in operation, but it protects you against major loss in the event of a hard disk failure. It also makes it convenient to work on other machines and in other places if the current version of your work is on diskette. Incidentally, since 1984 I have had only two instances of diskette corruption. Perversely, one of these was during the writing of this chapter! Fortunately, I was able to retrieve all except one page of text from the backups. Backing up your data is boring, but it is a small insurance premium against losing months, or even years, of work. Take it seriously!

7.3 IT in the enterprise

Now consider how IT is used in the business enterprise as a whole rather than by just you in support of your immediate work. We have stressed the importance of the business information system which embraces people, procedures and technology. Within that total system the technology element will be the corporate IT system. It probably includes computer hardware, software and telecommunications. If you are hazy about the scope of these, you might find it useful to look at the brief overview given in Section 7.1, or at one of the references mentioned there.

Consider the general case: a multinational company which has a large corporate IT system. Each site will have devices like PCs, printers and telephones linked by a local area network. These in turn are connected by a wide area network which will use national and international transmission and switching facilities. The system will be multimedia. For example, the computer manufacturer Digital has a corporate IT system which includes some 85,000 nodes. It offers electronic mail (e-mail), videotext for data retrieval, computer conferencing and video-conferencing. Further details of this and other international systems are given in Silk (1995). Here, I discuss practical aspects of using such systems, not their technical detail.

It will help to classify the various types of IT-supported business communication system in Figure 7.2. The first distinction to draw is between messaging systems

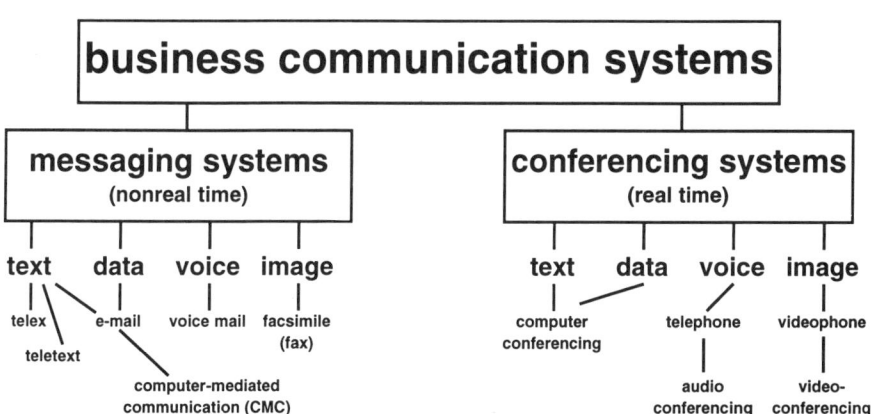

Figure 7.2   Business communication systems

and conferencing systems. In a messaging system there is no need for the sender and the receiver to participate at the same time. The sender transmits a message which goes electronically through the system. The message is then held in storage until the recipient is ready to accept it. The storage can be electronic or in hard copy (printed) form. The key feature of a messaging system is that the recipient can choose when to retrieve the message, study it and take any action. The action might include sending new messages, such as one back to the sender of the first message. A conferencing system is quite different. In that case all the participants (there will be at least two of them) must be available at the same time. They interact over the IT system. Normally each participant can receive from, and transmit to, all the others simultaneously. This is sometimes called a real-time conferencing system, to make it quite clear. The participants are in effect conducting a conversation, in whichever mode the system supports.

The next level of classification in Figure 7.2, for both messaging and conferencing systems, is the mode of communication. There are four main modes of handling information: text, data, voice and image. In business we need all four. Figure 7.2 shows the main types of business communication system for each of the eight categories.

Think about this classification in the context of your own work situation. What are the relative merits of messaging and conferencing systems? Does this relate to the distinction between urgent business and important business? Which of the four modes is suited to particular purposes? How many of the example systems have you used, and what conclusions have you drawn about their value? Think about that before we move on.

Now look at the practical aspects of using these systems in business, taking the messaging systems first. Telex is the most basic of all; it is a long-established method of communication which sends messages between teleprinters. It is often the only reliable method of text communication to remote parts of the world. Messages should be short. They are usually constrained to capital letters, like an old telegram. Telex is fine for sending short, factual information. It is difficult to make it sound personal. Today, it is a necessary standby to use when other systems are not available.

Teletex is a more modern alternative. It is intended to link PC-based wordprocessors, over a telephone connection (PSTN or ISDN). Although mainly for text documents, the more advanced standards enable you to incorporate data files and still images into the document which is transmitted. Teletex has not been adopted as widely as was expected in the early 1980s, largely because of the growth of electronic mail, or e-mail.

E-mail is now the main text-mode messaging system. It is offered on private, corporate systems and on public networks such as Internet. You can generate messages on a PC or similar terminal and send it to one or more addressees. In some cases you can specify the priority of the message; it is important not to abuse this feature or everybody suffers. You may be able to encrypt (scramble) the text of the message so that it can only be read by the intended recipient; this is a useful feature for sensitive or confidential information. You may be able to digitally 'sign' the document; this is evidence that it was sent by you personally and is a guard against forgery of messages.

Most people say that e-mail fills a gap in the range of business communication systems. It is less formal than a written letter, but more formal than an ephemeral telephone conversation. You can be a little more personal and even chatty. However, there are several points to watch:

- As a user of an e-mail system make sure you log-on regularly. In the office you may be connected all the time. For other systems, or away from the office, log-on regularly (perhaps daily). It is very frustrating for everyone if people are not available for lengthy periods; it defeats one of the main advantages of e-mail.

- It is very easy to send messages to far more people than need them. This results in electronic junk mail which wastes a lot of time. Think carefully about who needs what.
- Remember that if you are replying to a message from someone else they will not see it for some time, long after they sent the original message. Therefore do not respond in a hasty or even angry way; it may seem incongruous to the other person. One of the strengths of e-mail is that you can give careful consideration before you respond.
- Avoid the instinct to reply to everything you receive.
- Be informal, but be careful about humour. If the recipient is in a less than cheerful mood, your witty remarks in text on the screen may not have the effect that you intend. Some e-mail addicts include combinations of standard symbols to indicate emotions:

    :-) I'm happy
    :-(I'm not happy
    :* I'm drunk

    Turn your head to the left to understand those. Incidentally, it is generally considered rude to USE CAPITAL LETTERS in the text of an e-mail message; it it the equivalent of shouting in conversation.

An extension of e-mail is computer-mediated communication (CMC). A central computer holds a database of contributions on particular topics. Those interested in a topic can read what has been said earlier, and add their own contributions. This leads to a communal, offline discussion in the text mode. Systems also allow you to append files of data, but the principal mode is text. For CMC you need the same disciplines as for an e-mail discussion with one other person, but you now have the added complexity of a group discussion. You are in effect taking part in a nonreal time meeting. The roles of individuals resemble those in face-to-face meetings; these are discussed in Section 6.3. In particular, it is helpful to have a leader or chair of a discussion on each topic. Modern CMC products are called groupware; their importance is mentioned in the following section.

E-mail and CMC operate in the text mode. Voice mail or voice messaging is the equivalent in the voice mode. It is a facility normally associated with the telephone exchange on a particular site. People ringing in can leave a recorded message if their intended recipient is not available to answer the phone. When that person returns and picks up their phone they are alerted to the fact that there is a voice message awaiting them. Using a code on the keypad enables them to listen to the message.

Voice mail is like a communal telephone answering machine, with extra facilities. As with answering machines, there are people who don't like leaving recorded messages. However, it is best to develop the skill as voice mail is expanding rapidly. Think about the core of your message so that you can express it succinctly if you get the voice mail instead of the person you expected to speak to. In any case, you have a few seconds' warning before you have to start speaking. It helps to give your name and the day/time that you are making the call, as well as your return phone number. Even if the recipient knows your number, this makes it easier for them to call you back after they have listened to your message. Take great care to give such details slowly and clearly. Repeat them, even. It is very easy for you to recite your phone number quickly; it is much more difficult for the other person to note it down. I sometimes have to replay messages three times in order to get a phone number which has been gabbled too quickly.

Some voice mail systems enable you to send the same recorded message to more than one recipient. As with e-mail there is a severe danger of creating junk mail and a less personal style of communication; be careful! However, both e-mail and voice mail are useful for overcoming the barriers of time-zone differences; this is in the nature of messaging systems.

A similar advantage attaches to facsimile (fax). This enables a copy of an A4-sized document to be sent over a telephone connection. Over the PSTN a page usually takes about one minute to send; over the ISDN it is at least three times faster. In both cases fax is not only instantaneous but is cheaper than mail for anything less than about four sheets.

Fax machines are now cheap enough to buy as a personal, desk-top device. However, a fax capability is often built into office automation systems so that all transactions can be done from a single PC-like terminal. Those systems offer automatic multiple addressing. Some also include scanning of incoming text messages, to convert them into a computer file which is compatible with a wordprocessing system; this can save a lot of rekeying time.

The use of fax is expanding rapidly, aided by improved international communications and the wide adoption of the technical standards. Rather like e-mail, it is somewhere between a letter and a phone call in terms of formality. A lot of business is now done by fax; make sure you are aware of the legal status of contracts entered into during a fax dialogue. Sometimes it is a convention in a particular industry that a fax message will be accepted as a legally binding order, for example.

Be cautious about giving your fax number to people who produce directories. You are likely to be the recipient of a large number of unwanted (junk) faxes as a result. They use up your fax paper and prevent your fax being accessible by those who need to contact you. Most fax machines, whether individual or corporate, add a header which shows where the fax has come from (name and number). Some businesses are very sloppy about this identification line; make sure it is appropriate and helpful. Also remember to change the time during the daylight-saving months. It helps to have a cover sheet for every fax showing the originator, contact details, the number of pages, and the return fax and telephone numbers. Faxes are quick, cheap and convenient. They often produce prompter action than phone calls or letters.

Let's consider the real-time conferencing systems shown in Figure 7.2. The first is computer conferencing. Like e-mail the mode is basically text from a keyboard to a screen, but this time all the participants take part at once. If there are more than two participants one of them needs to chair or co-ordinate the session. Generally this system is less effective than computer-mediated communication (see earlier). However, it can be useful for one-to-one communication. It is important, to consider each of your contributions to the dialogue carefully. Trying to maintain a quick-fire dialogue is not appropriate, especially if you are not both good touch typists. It leads to frustration, and you would be better off using the telephone.

In the voice mode, the telephone is of course the most familiar device. Because we are so familiar with it we tend to forget how rude and intrusive it can be. Gradually, it may become good manners (as well as more convenient) to use a voice mail or other messaging system whenever you can. There are some other points to remember about using the telephone in business:

— Often the telephone is the first contact which customers and other outsiders have with your firm. Make sure that the people who deal with calls regularly (whether on the switchboard or in departments of the business) are trained. A good telephone manner can help the customer, and give the impression of an efficient and caring organisation.
— Most telephone calls are less important then the work which they interrupt. Consider whether you can use a messaging system instead, especially if the topic requires careful thought rather than an instant response.
— When you ring someone, you already have a picture of them in your mind and have been thinking about the problem you wish to discuss. When they answer the phone, they are probably thinking about something entirely different. So give a greeting and your name. Allow a moment for this to sink in. Then go on with the conversation. It is rude to assume that people will recognise you by your voice alone. It is most unhelpful to launch straight into the detail of what you have to say.
— Don't be lured into hasty responses. If you need to check some facts, or think about the issues, say so. If you promise to ring back, say when and then make sure you keep the promise.
— The telephone is a means of having a conversation at a distance. It is good to exchange socially pleasantries, but make them short and sweet.

The audio conference links more than two people by telephone simultaneously. They can be quite successful, better for example than real-time text-mode computer conferencing. Again, it is best to have a chair or co-ordinator. If the topic warrants it, a written note of what was agreed should be prepared.

The telephone and the audio conference are cheap and widely available. They can be personal, but the inability to see the other participant(s) limits the degree of empathy. You are denied most of the cues of nonverbal communication discussed in Section 2.2. IT now offers affordable solutions to this problem.

The first is the videophone, intended as a desk-top device for use between individuals. Be careful about these. There are proprietary videophones which work (more or less) over ordinary analogue (PSTN) telephone connections. However, the picture size and quality are very limited and they are not compatible with anything else. They also degrade the speech quality. You might buy a pair to talk to granny. For business use there is not yet a general standard for videophones, except for the digital (ISDN) network. For the ISDN there is now a second generation of videophone and videoconferencing technology. This works over two simultaneous ISDN telephone connections, so the running cost is comparable with a phone call. The quality is adequate for most business purposes, giving user recognition and the ability to look at diagrams and engineering models. The terminal may be installed in a special room or studio, or be on the individual desk-top as a separate device or integrated with a PC.

There are some problems in using videophones. Do you always wish to be accessible visually and well as aurally in your work location? If you turn off the camera, does this also send a message to people who ring you? Try to maintain apparent eye contact, because that is an important benefit of videophones. However, to do this you must remember to look at the camera rather than at the face of the other person displayed on the screen. If you forget, and do the natural thing by looking at the image, at the other end you will give the appearance of avoiding eye contact. You might even look shifty!

Videoconferencing is the extension of this technology to involve more than two people. The arrangement attempts to create the impression of a face-to-face meeting. Here are some points to consider:

- The first generation of videoconferencing (from the early 1980s) used high speed digital connections. It has proved itself very useful for meetings, especially between people like designers who need to exchange and look at drawings and hardware. The equipment is still valuable for that.
- The new ISDN-based equipment offers a much cheaper alternative for remote meetings. There you should try to create the impression of sitting around a 'virtual meeting-table', so that the psychology of a normal meeting is nurtured (see Chapter 6).
- The aim is to make the technology incidental to the proceedings; the less aware people are of it the more successful it has been.
- You must employ the usual disciplines of running business meetings.
- Videoconferencing does not replace face-to-face contact between colleagues and customers, but it does mean that such contact can be less frequent. Short encounters are also much easier to arrange. In many organisations there is a good business case (based on saving of time and travel expenses) for using the new range of equipment.

That concludes the survey of using modern IT-based business communication systems. If you want further material on this topic, see Peel (1990). Each system has its own strengths and weaknesses, but you can get the most out of each if you give some thought to the person at the other end and the limitations of the medium. It is like using a satellite voice link; you must learn to cope with the time delay if you don't want to trip up in conversation. The technology is becoming more fully integrated, so that you have a few multifunctional devices on your desk rather than an array of single-function devices. Within the network the trend is to integrate into a single digital system, notably the ISDN. As this wider choice becomes available, you may find the technology becomes less intrusive and limiting. You will then be able to replicate the way we deal with colleagues alongside us. This might mean video links to enable conversational exchanges in a conferencing system, and multimedia documents for detailed or technical exchanges in a messaging system. Does that sound to you like the way we ought to go?

7.4 The future

The preceding section ended with a speculation about how we will use IT as it becomes more powerful than it is now. There is no doubt that it will become more powerful; we are nowhere near the fundamental limits for IT development. This section looks at some of the changes which are currently happening in business and which set trends we can expect to continue: dispersed teamwork; teleworking; project-oriented management; corporate interdependence; the emergence of the flexible, learning organisation; and the growth of the Internet. Some of these topics link with the management trends identified in Section 1.2, but here the focus is on those which are made possible by modern IT systems. An engineer in the modern world of business should consider these issues carefully.

Teamwork was discussed in Section 6.4, in the context of business meetings. The preceding section looked at computer-mediated communication (CMC) which enables teams which are dispersed geographically to interact more easily. The software which facilitates this is called groupware. By the mid 1990s many firms were using groupware (such as Lotus Notes) extensively. If your work involves working closely with people at a distance, over a protracted period, then groupware could help. What do you need to bear in mind when using groupware? Here are some practical points:

— Because of the relative informality of a groupware conference the style tends to be slightly different than in a face-to-face business meeting. The group tends to work as a peer group even though there may well be an organisational hierarchy which is understood by all those involved.
— Logic and persuasion are therefore more effective than authority and precedent. This can bring a refreshing perspective to discussion. It can also slow things up. The group needs to develop its own working procedures, as in the second ('storming') stage of team development discussed in Section 6.4.
— There is no reason why the dispersed group should not develop to the fourth ('performing') stage. This may be possible even if the members never meet face-to-face. They then become a virtual team, perhaps.
— Despite that, it usually helps if the group members can meet periodically. Research shows that electronic interaction reduces the frequency with which face-to-face meetings are needed to maintain the cohesion of the group.

An obvious extension of dispersed working enabled by IT systems is the idea of teleworking. This is where individuals spend more than half their time working from home, or another remote location, using IT to maintain their contacts with the parent organisation. There are attractions for both parties. The teleworker enjoys a more flexible lifestyle without the hassle of commuting. The firm saves on overheads and can pay for results rather than for attendance. However, teleworking requires discipline on both sides. If you are teleworking you need considerable self discipline with regard to your working habits. You must also be prepared to forego some of the social contact which is enjoyed in the workplace. You might decide to visit regularly just to keep in touch with the organisation and its informal networks. Some firms provide videophones, to help keep the personal touch. If you are a manager of teleworkers, you must learn to judge by results produced, rather than the way in which the teleworker chooses to operate. In particular, you must not judge by time spent at the electronic workplace. Teleworking is on the increase, and will continue to do so as the nature of work and individual careers change. We must learn to have a much more flexible relationship with the organisations we work for; see Handy (1994).

A management change which is related to this is the greater emphasis on project-based management. A project is an endeavour in which resources are organised to undertake a unique, specified task within cost and time constraints. Increasingly, business management is less about 'business as usual' and more about achieving a series of planned changes in a controlled manner, in other words project management. Projects require expertise to be brought to bear flexibly over a defined period of time. Dispersed working, supported by the use of groupware, can be a great help to this. Indeed the project data can often be shared by supplier and customer, whether they work in the same or different organisations.

This is one example of a broader trend: towards greater interdependence between business organisations. The IT systems enable most information to be shared, leading to greater openness and a closer relationship at the working level. Most engineers who have experienced the frustrations of formal, arms-length relationships in projects will welcome that trend. It is a trend which has been identified as a major one in the business life of the 1990s. The attitude should be one of long-term partnership with other businesses, rather than of locking-in or exploiting them. At the detailed level of business communication this therefore means openness, courtesy and a mutual concern to get the job done to the benefit of both parties. Some of the old attitudes have to change, and perhaps you will have a role in that.

Business commentators sum these trends up by describing the flexible, learning organisation. The organisation must be knowledge-based, and bring its resources to bear flexibly. IT systems have a part to play in this at several levels. All human endeavours, apart from that of the hermit, rely on human communication. Business communication, underpinned by the use of modern IT systems, therefore has a central role in the development of the new style of business organisation.

One of the most striking examples of an IT resource which has suddenly had an impact on business is the Internet. The Internet is not a single resource run by a single authority. It is a connected system of individual computer networks around the world. These individual systems work together by using standard technical protocols. Thus Internet is an example of how technical standards can lead to useful international systems; another example is the impact of the standards for fax machines. Internet started in 1969 as a prototype US defence system. By 1983 the standards were stable, and this lead to a rapid expansion of the system. In 1989 the first public Internet servers were connected. This lead to a rapid expansion. No-one knows exactly how many users are connected to Internet; by 1995 it was estimated at 20–30 million. By 1996 more than half would be outside the USA.

Internet is not just a communication system. It is overlaid by services which enable users to store and share information. Examples are Usenet, World Wide Web ('The Web') and multimedia services. In 1994, after several years' use by technical enthusiasts, Internet attracted its first serious commercial users. The transition of the Internet to become a universal international IT system is fascinating. It could have a major effect on national and international business. Many people view it as providing the information trade routes for the 1990s and beyond.

That concludes a very brief survey of some of the IT-enabled changes which are already under way. We can expect them to accelerate. Engineers are better placed than many other groups of people to understand these developments and learn to take advantage of them. That means keeping abreast of developments and applying the promising ones with an open mind.

7.5 Summary

This chapter has considered business communication from a different perspective: as a transaction within the total business information system rather than as one step in solving a management problem. The business information system comprises people, procedures and technology. Its function is to collect, store, process and communicate information to support management decision and action. It must handle both internal and external flows of information. The latter, particularly, are often informal and longer term in their relevance.

To set the scene, information technology and its current trends were briefly considered. Here are the main points:

- IT is the convergence of computing and communications made possible by modern microelectronics.
- The power of integrated circuits (ICs, or chips) has developed at about 30% per year since 1960. Computer hardware has developed in step with the power of ICs, giving greater power, smaller size and lower cost. The laptop/notepad computer exemplifies this.
- In turn, this supports more complex computer software. Simple applications are integrated, so that they can share data; the user interface is becoming more

convenient; PCs are networked within the work environment; client-server technology and groupware support new co-operative forms of working.

- Communications transmission has been enhanced with fibre-optic cable. This has released radio capacity for mobile communication systems.
- Wide area networks such as the public switched telephone network usually have an analogue user interface. Modems can be used for digital communication, but wholly digital networks are better. The integrated services digital network is the major system for business, and is now available internationally.
- As a user, you should be aware of two legal aspects of using IT: data protection, and computer misuse. Both have supporting legislation.

Section 7.2 invited you to consider your own information system: the way you collect, store, process and communicate information to carry out your job. You need to consider this as an information system before you can decide the appropriate role of IT to support it. You should also consider what IT is available to you and the administrative support which you get. In most jobs it is the personal computer, or equivalent network terminal, which offers the obvious starting-point for using IT as a personal aid. Consider word processing, spreadsheets, graphics, desk-top publishing and specialised statistical packages. For most people, word processing is the most rewarding place to start. Remember the practical advice about touch typing, spell checkers, grammar checkers and keeping regular backups of all your work. The PC can be a liberating tool, but it requires some practical discipline from the user.

Section 7.3 looked at the role of IT in the wider business enterprise. The corporate IT system may be very complex, so we used a classification of the main types of business communication system (Figure 7.2). The practical aspects of using messaging (nonreal-time) systems were discussed: telex, teletex, e-mail, computer-mediated communication (CMC), voice mail and fax, then conferencing (real-time) systems: computer conferencing, the telephone, audio conferencing, the videophone and videoconferencing. The trend is towards multimedia, offering you a wider choice of ways to communicate.

Finally, Section 7.4 looked at some of the likely trends for the future. Mentioned briefly were dispersed teams supported by groupware; the growth of teleworking; the greater emphasis on project-based management; increased corporate interdependence; and the development of the flexible, learning organisation. Internet is a remarkable example of an IT resource whose business significance has only recently been recognised.

IT systems have a key role in the development of businesses, as well as in support of individuals who work in those businesses. You may so far have experienced only a few of the systems mentioned. But one thing is certain: we shall all be exposed to more of them in the future. Let's make sure we are ready for that. Remember that whoever wrote the item quoted at the start of this chapter thought they were giving a fair assessment of the likely business impact of IT. Make sure that you do not similarly misjudge!

Chapter 8
Conclusion

'Information is expensive, but lack of it is more so'
Lord Rayleigh (1842–1919)
'He who considers everything decides nothing'
Chinese proverb
'Only a word of command, but it loses or wins the field;
Only a stroke of the pen, but a heart is broken or healed'
F R Havergal (1836–1879)

8.1 What's the message?

As you know, every good business communication ends with a summary of the main points. For this book, it is in this section. Figure 8.1 is a reminder of the structure of the book, which was explained back in Section 1.3. It is a copy of Figure 1.2. After the introductory chapter, each of the main chapters had its own summarising conclusion. This section simply reminds you of the main points covered. If you need to refresh your memory, look back at the summary at the end of the chapter concerned. If necessary, go back to the detail within that chapter. If you have not yet studied all the main chapters then the summary here may suggest topics that you might wish to look at next. It was emphasised that after Chapters 1 and 2 the book could be read flexibly; however, the end of this section offers some tips about developing your skills systematically.

Chapter 2 (Communication and management) provides the framework for the detailed chapters. It offers a model to understand the various methods of human communication. The key features are transmission and feedback, within the shared

143

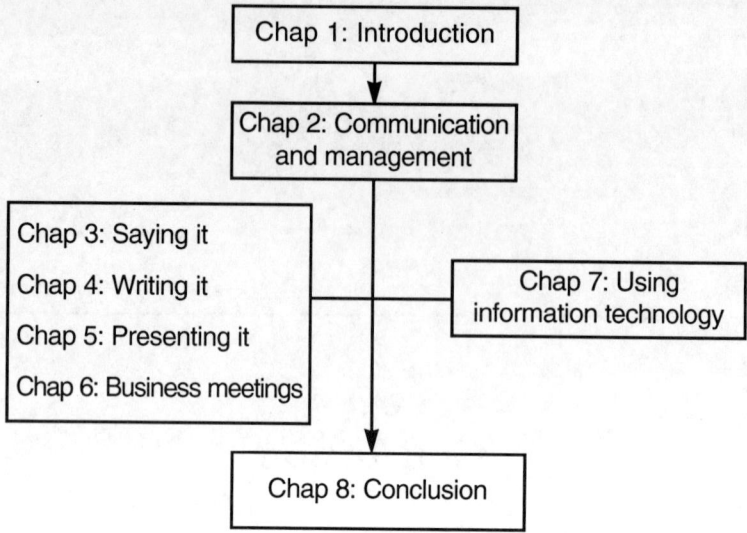

Figure 8.1   Structure of the book

context of sender and receiver. Speaking and writing are considered in two
contrasting situations: one-to-one/few and one-to-many. The limitations of human
information processing show that we need redundancy (repetition) in the message
if it is to be understood clearly. Chapter 2 then looks at trends in business. The
wider scope of business relationships and the greater demands of the market mean
that business communication is becoming more important than ever. To keep
pace, every manager is engaged in a learning cycle (do–watch–think–try). This
is underpinned by an information value chain (data–information–decision–action).
This model helps us to define the purpose, aim and scope of each process of business
communication.

   Chapter 3 (Saying it) looks at the most common method of human communication. Within the business enterprise we need to talk with a wide range
of people; some may be peer group colleagues, others internal customers of the
department we work in. Active listening is a special skill which makes spoken
communication more effective. Transactional analysis helps us to understand
conversation, and deal with possible sources of conflict. As with all business
activities, we need to consider both the task and the process. Engineers tend to
be task oriented, and need to cultivate a greater awareness of process. When
speaking to people outside our business organisation, the challenge can be even
greater. We must consider issues of business culture and national culture. There
are differences of custom, especially for nonverbal communication, which we must
respect. There are also special problems in dealing with the media and with making
speeches. For the words themselves, language is a powerful tool with some serious
limitations. We must be aware of the pitfalls of ambiguity, emotion and logical
error. In business communication, be logical where you can; be persuasive where
you can't.

Chapter 4 (Writing it) considers several types of written communication in business. Besides using words, we may need to use figures and diagrams to present quantitative data in a meaningful way. Chapter 4 focuses on the most formal and structured type of business communication: the management report. Two principles add to those identified earlier: offer the reader multiple levels of detail; and separate fact, analysis and opinion. A structure is offered as a template to meet these requirements. This completes the acronym PASS (purpose, aim, scope and structure) needed to plan any good business communication. A practical nine-step procedure for drafting reports is explained; it can be modified in the light of your own preference and experience.

Chapter 5 (Presenting it) deals with a quite different type of formal business communication: presentations. They are often the best opportunity to influence senior decision makers. Besides planning the presentation itself, probably as a team effort, you must deal with a wide range of practical issues. These include venue, timing, use of notes, visual aids, other technological support, and possibly handouts. Several rehearsals are needed, if possible in the final venue. On the occasion itself, you should check the preparations, receive the audience, deliver the presentation itself and then deal with questions. Lead smoothly into the next part of their business. Make sure you review the effectiveness of the whole process afterwards.

Chapter 6 (Business meetings) uses the information value chain to identify the scope and shape of a business meeting. This determines who should, and who should not, attend. As for a presentation, planning is vital. The additional factors are the formal agenda, and the jobs of chair, secretary, scribe and timekeeper. The type of formal output must be decided: action list, brief minutes, or full minutes. Before the meeting you need to greet those attending; this requires social grace and the ability to remember key names. During the major items of the agenda, the chair has a vital role in controlling the discussion. The chair should introduce and conclude each main discussion. In between, there needs to be a diverge–converge pattern. The diverge phase is relatively easy, with tools like brainstorming available. The converge phase is more difficult as it requires people to adapt their views. A skillful chair will see a possible route through the maze, to offer to the meeting if necessary. Normally the group should aim for consensus agreement, with full commitment. Voting is a poor alternative; unanimity is usually an unrealistic goal.

Chapter 7 (Using information technology) considers a business communication from a different perspective: as a transaction within the business information system, rather than as a step in a management process. There is a brief overview of modern IT developments and the related legal areas of data protection and computer misuse. Chapter 7 then makes some suggestions about your personal information system, and how you might use IT to improve it. The PC is an important tool there, and most people start with word processing. However, you must be realistic about the time and effort needed to become effective. This relates, in turn, to the corporate information system and the practical aspects of using modern IT systems, both messaging (nonreal-time) and conferencing (real-time).

Finally, Chapter 7 looks at some current business trends which are underpinned by the rapid development of these IT systems.

That completes a very brief summary of the content of the main chapters of this book. Stop for a moment and consider how you are feeling. The aim of the book is to help engineers to communicate effectively with nonengineers, in the business context. Do you feel that together we have yet achieved this aim? There are four possibilities:

1   You are still feeling somewhat bemused by what you have read. In that case, go back to the summary at the end of the appropriate chapter. If necessary reread the relevant parts of the chapter itself. That gives you access to several levels of detail.
2   You have completed and understood what you planned to read, but now think that you would like to look at some of the other topics as well. Provided you have digested Chapter 2, you can go straight to the relevant detailed chapter. It includes a summary at the end, of course.
3   You have understood the material which you want to study from the book, but feel you would like to examine some of the topics in greater depth. In that case, look at the References and further reading which follow this chapter. All of them (except reference books) have been mentioned at appropriate points in the main text. However, brief guidance is given under each reference to help you choose what will help you most. There are two general books which might help you: Peel (1990) and Stanton (1986).
4   The fourth possibility is that you have understood what you have studied, and now want to improve your performance in your particular work situation.

Ideally, you should go back through the earlier material, and look at additional references, until you feel that the last statement accurately describes you!

What then? The first thing is to draw up a list of your development needs. This is managementspeak for 'weaknesses', so make sure you are honest about them. Consider them carefully, under the headings of Chapters 3–7 of the book: speaking, writing, presenting, running meetings, and using IT. Here are some examples of development needs, to help you assess your own situation

Speaking:
    'I use jargon too much. Nonengineers switch off when I'm talking'
    'I'm OK on technical subjects, but can't talk about the business'
    'I need to listen more carefully to what other people are saying'
    'I upset people by being too impatient'
    'I need to understand cultural differences better'
    'I need to be more logical and persuasive'

Writing:
    'I find it difficult to get my ideas down on paper'
    'I can write technical reports but not management reports'
    'I find it hard to distil quantitative data into summary form'
    'I need to get more structure into my reports; they aren't helpful'

'I need a systematic procedure for drafting reports'
'I always run out of time, and produce a rushed report'

Presenting:

'I'm scared stiff of giving a presentation'
'They all look bored when I'm talking'
'I wish I could use speaking notes unobtrusively'
'My visual-aids always seem too cluttered with detail'
'We always seem to get the timing wrong'
'The presentation goes well, but we trip up on the questions'

Running meetings:

'We have mammoth meetings where most people never say anything'
'I need to delegate some of the jobs of running a meeting'
'I find it difficult to be the chair or secretary of a meeting'
'Our meetings run over time, and people get restive and annoyed'
'It's never clear who should do what after the meeting'
'People say they will do things, and then don't'

Using IT:

'I need to consider using IT more in my work'
'I need to check whether we comply with legal requirements'
'I use PCs, but in a rather disorganised way'
'I must find out what the corporate IT system offers me'
'Perhaps IT could change my workstyle for the better'
'New technology could change our business; I must find out more'

Another way to help you identify development needs is to perform a self assessment. Figure 8.2 gives an approach to that. It is a modified form of the strengths,

strengths	weaknesses	
1 strengths + opportunities (GOOD)	2 weaknesses + opportunities	opportunities
3 strengths + threats	4 weaknesses + threats (BAD)	threats

Figure 8.2   Self assessment

weaknesses, opportunities and threats (SWOT) analysis often used in business. First identify your strengths and weaknesses in the area of business communication. This book should have helped you to identify them clearly. Write them in the boxes on the upper edge of the matrix. Next identify the threats and opportunities which your business situation presents. This might be more difficult. An example of an opportunity is the prospect of promotion which would involve much more business communication. An example of a threat might be the introduction of a new IT system which you are going to have to learn to use. Write the opportunities and threats in the boxes on the right-hand side of the matrix.

Now consider the four boxes in the main part of the matrix, to identify your development needs. For example, an opportunity may require you to build upon a strength (box 1), or correct a perceived weakness (box 2). A threat is more likely to require you to correct a weakness (box 4). In this way you can identify development needs which

- enhance your ability to seize opportunities
- reduce your vulnerability to threats

Armed with your carefully considered list of development needs, the next stage is to prepare an action plan. An approach is shown in Figure 8.3.

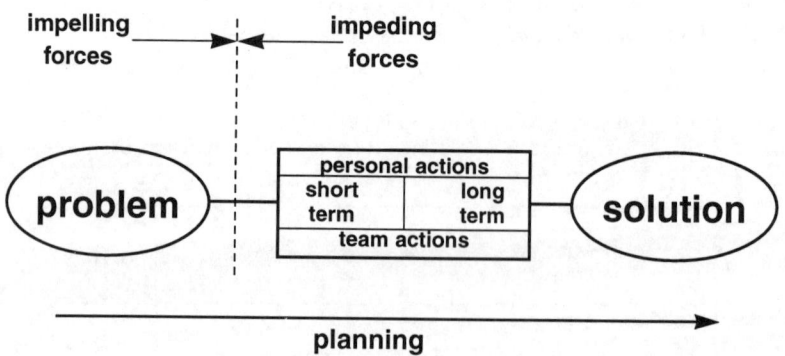

Figure 8.3   Action planning

For each of your development needs, consider carefully what are the
1   impelling forces: the factors which make a solution to the problem easier
2   impeding forces: the factors which make a solution to the problem more difficult

Now look at these forces more closely, and identify

3   which impelling forces you can take advantage of to help meet the development need
4   Which impeding forces you need to counter or avoid to meet your development need

This will in turn identify

5   personal actions which you must take yourself
6   team actions where you will need to share the action with others

In both cases it may be useful to distinguish between short-term and long-term actions.

This approach is illustrated with one of the development needs listed earlier: 'I find it difficult to be the chair or secretary of a meeting'. The major impelling force might be the fact that your boss chairs a regular series of meetings to review progress of a project which you are managing. Well, at least there are meetings which you regularly attend! You might identify the major impeding force as the fact that your boss is reluctant to delegate and has always chaired the meetings personally and used another member of staff to act as secretary. This is a case where you can solve the problem by removing an impeding force rather than increasing an impelling force. That is quite often the case. So, in step 4 you decide you must tackle the boss about the problem. In step 5 you decide your approach to this. You might tell the boss you feel you should be developing your skill in running meetings. The boss can hardly quibble with that! Then you suggest that you might take on the role of secretary of the routine meetings. This would be a short-term action, which you might have agreed in principle with the person who is currently doing the job. They are unlikely to raise much objection. Assuming this goes well, and you prove an effective secretary, you might suggest to the boss that you could well chair the proceedings. This would free the boss from the routine task, although he or she could still attend and sit on the sidelines. A thoughtful boss would find it difficult to resist this approach, and might even be quite impressed.

Your action plan involves you in extra work, but that is in the nature of self development. However, it has been made possible by removing obstacles rather than by forcing your way through the problem.

Think about your own development needs in the light of this approach, and then draw up your own action list.

8.2 What next?

Review your development needs and action plan regularly. You might even need to refer back to this book or some of the suggested additional reading. Developing your business communication skills should be a long-term project. You can never sit back and say 'There, I've done it'. But you should be able to say 'Well, I've made a distinct improvement in the areas I planned to deal with; now I'll tackle some others'. In this way you will contribute in an important way to your own career progression. Recall that Section 1.2 identified three stages to a typical career: engineer, manager of engineers, and general manager. However far you plan to go along that progression, better communication skills will ensure you do a better job.

Sooner or later you find yourself with a management responsibity for other people. That does not just mean being responsible for the immediate work they do. It also means taking care of them as people in the organisation: nurturing their skills and helping them to become more valuable members of the organisation. One way of interpreting this is to consider that you have a responsibility to develop your subordinates so that they can do your job. One aspect of that is to help them to develop their professional and business skills. Today, that is seen as a joint responsibility between the employee and the employer. When you are in a management position, you are an employee as far as your own development is concerned, and you are the employer as far as the development of your staff is concerned.

Think about the boss in the case we discussed at the end of the last section. There, you (the employee) were left to analyse your needs and decide on an action plan. This book may have helped you. But how much better if the boss had been more proactive. Perhaps as part of the routine appraisal system the boss should discuss your development needs and how, jointly, you can meet them. Business communication skills would be considered alongside professional and other business skills.

Some organisations and bosses prescribe what training you should do, with little attention to your particular needs. Others leave it to you to make a fuss, and then they might help. The more enlightened ones will enter a partnership for your development. It is, after all, in your mutual interest that this is successful.

So as a conscientious engineer and manager in business, make sure that you look after the development of your staff, as well as the development of yourself!

A final word about business communication skills. You need to set yourself performance goals to aim for. Here are some which were produced by the man to whom this book is dedicated, David Ellis-Jones (1919–1992). They are quoted with the kind permission of his widow, and define the standards which college and university lecturers should set themselves:

- I will have prepared a learning experience
- I will not be overstructured
- I will make listening easy
- I will ensure my visual aids are for my listeners
- I will be relaxed and devoid of tension
- I will at the end of my task reinforce the vital content
- I will remember that if I emphasise everything I emphasise nothing
- I will not have too much content and rush to fit it in
- I will be proud of my product, on which the College will be judged
- I will be professional
- I aim for a feedback grade of 5 [the highest]; I will be content with 4; anything less will tell me I need help.

Those are wise words which can be applied to most types of business communication. They sum up what this book has tried to say. Let's all try to put them into practice.

References and further reading

This list is in alphabetical order of author. In the case of reference works such as dictionaries, the title is shown instead. The short abstract will help you choose items for further reading. The term 'here' refers to 'How to communicate in business', the book you are now holding.

ARGYLE,, Michael (1988): 'Bodily communication' (Methuen, London, 2nd edn.)
Argyle approaches the topic as a social psychologist, giving a scholarly overview but in a nontechnical and engaging style. Nonverbal communication is socially very important and the book is therefore of general interest. However, there is also specific guidance relevant to management communication, including for example cross-cultural differences in the meaning of gesture. An authoritative book on the subject. It relates to Section 3.3 here. See also Morris (1994).

BARLOW, Horace *et al.* (Eds.) (1990): 'Images and understanding' (Cambridge University Press, Cambridge, UK)
Comprehensive and authoritative, this collection of essays needs close attention but gives rewarding insights. Besides dealing with the brain and perception, it considers nonverbal communication, visual art, and the interpretation of scientific images. Read this if you are fascinated by how images achieve their effect. See Section 2.2 here.

BELBIN, Meredith R. (1984): 'Management teams: why they succeed or fail'. (Butterworth–Heinemann, Oxford, UK)

BELBIN, Meredith R. (1993): 'Team roles at work' (Butterworth–Heinemann, Oxford, UK)
Dr. Meredith Belbin conducted the classic research on the roles which need to be performed in an effective management team. There are eight general team roles and one special one. They are listed in Section 6.3 here, and have entered general management parlance. Belbin's 1984 book is a management classic, with principles which will be of value to anyone responsible for teams at work. The 1993 book adds the experience of applying the principles in a wide range of contexts.

BERNE, Eric (1966): 'Games people play: the psychology of human relationships' (Deutsch, London, UK, also Penguin, London, UK, 1968)

BERNE, Eric (1974): 'What do you say after you say hello?' (Deutsch, London, UK. Also published by Corgi, London, UK, 1975)

151

These two books include a description of transactional analysis (TA). They therefore develop the detail given here in Section 3.2. The Parent, Adult, Child model is used to analyse a wide range of social interactions, whereas here it has been used only in the context of business interactions. The books give a fascinating framework for understand human behaviour and interactions.

BUZAN, Tony (1988): 'Make the most of your mind' (Pan Books, London, UK, revised edn.)

Buzan gives practical guidance about using the power of your brain to better effect. It is a book of general interest to managers, but is relevant to our topic because it gives empirical evidence about retention in the human memory. This has important implications for learning and for structuring messages, whether written or spoken, as discussed in Section 2.2 here. There are self-check exercises which help you understand the capability of your own memory. For example in the task of reading he suggests four stages: survey, preview, inview and review. The first three address successively closer levels of detail. In the review stage you check that you have absorbed what you need. This might involve taking notes or constructing a mind-map.

BUZAN, Tony (1989): 'Use your head' (BBC Books, London, UK, revised edn.)

This deservedly popular book is a good practical guide. It starts by putting the complexity of the brain into context, and then turns to how to develop skills of reading and memory. The mind map is Buzan's special tool; he discusses and illustrates its use in a variety of situations. Thoroughly recommended as an adjunct to Section 2.2 here.

CHERRY, Colin (1978): 'On human communication' (MIT Press, Cambridge, MA, USA and London, UK, 3rd edn.)

Colin Cherry was a leading expert on the subject. He died in 1979. This third edition of his authoritative book is accessible to the general reader, but engineers will particularly appreciate the scientific approach he uses. It moves smoothly from the technical aspects of communication theory through to the human and social features. The book develops Section 2.2 here.

'Concise Oxford Dictionary' (Oxford University Press, Oxford, UK, 1990, 9th edn.)

A good dictionary is essential to any writer. It should not be too big, or it is inconvenient to use and may encourage you to use unusual words which will obscure your meaning for most readers. The Concise Oxford Dictionary is one of several dictionaries which will help you write what you mean, clearly. The ninth edition is a significant development, appropriate to modern business and general needs.

CORNER, John, and HAWTHORN, Jeremy (Eds.) (1980): 'Communication studies: an introductory reader' (Arnold, London, UK)

A well structured set of essays from a wide range of contributors. Inevitably, there are contrasting views expressed. It is a stimulating collection for the general reader, or as background material for the serious student, developing Section 2.2 here. For a more academic introduction, see Fiske (1982).

'Effective business communication: a director's guide' (1994): (Institute of Directors, London, UK)

This 80-page booklet takes a senior management look at communications in the broad sense. It is about sustaining partnerships with the various stakeholders in of a business (suppliers, customers, employees, external authorities). This includes issues like human motivation, technological options, legal aspects, the relationship between stress and urgency, direct marketing and use of English in business letters. A useful set of essays on topics which complement Sections 1.2 and 2.3 here.

FISKE, John (1982): 'Introduction to communication studies' (Methuen, London, UK)

A good introductory text. It develops the basic models of communication (semiotic and process) and interprets them in a broad social context. There is particular emphasis on the techniques and role of the media, relating to Section 3.3 here. See also Corner and Hawthorn (1980).

FOWLER, Alan (1990): 'Negotiation: skills and strategies' (Institute of Personnel Management, London, UK)

The book defines negotiation as 'a process of interaction by which two or more parties who consider they need to be jointly involved in an outcome, but who initially have different objectives, seek by the use of argument and persuasion to resolve their differences in order to achieve a mutually acceptable solution'. It identifies seven principles of successful negotiation, and gives practical guidance to achieving them. It relates to Section 3.3 here.

'Fowler's modern English usage', by FOWLER, H. W., revised by Sir Ernest Gowers, 1965 (Oxford University Press, Oxford, UK, 2nd edn., also as Oxford reference paperback, 1983)

A classic work on the correct use of English. It was first published in 1926 and for two generations has been used by people who care about precision and good usage. Some of its precepts are rather dated now, and some people view them as pedantic. Nevertheless, its alphabetical arrangement helps you to answer particular questions very quickly. For example, if you are unsure whether it should be 'different to' or 'different from', or if someone accuses you of using a split infinitive, this is your best source of help. It must be said, however, that the first sentence of Fowler's introduction to the first edition (1926) is a good example of what not to write today.

'Gowers' plain words': revised by GREENBAUM, S., and WHITCUT, J., 1986 (HMSO, London, UK, 3rd edn., also Pelican, London, UK, 1987)

'The complete plain words' was published by Sir Ernest Gowers in 1954, combining two of his earlier books. Gowers was a distinguished civil servant and hoped to improve the bureaucratic and somewhat pompous style of writing common at the time. Most of the book is in a conventional format, which you can study or browse through with both benefit and enjoyment. There is a 70-page checklist of words which often cause difficulty. This is useful for finding the answer to a particular problem, but for that purpose 'Fowler's Modern English Usage' (q.v.) is more comprehensive.

GREGORY, Richard L. (1990): 'Eye and brain: the psychology of seeing' (Weidenfeld and Nicholson, London, UK, 4th edn.)

An excellent nontechnical introduction to the topic by an expert on visual perception, developing Section 2.2 here. Starting with the physics of light, it covers the role of the eye and the brain in the perception of images. It discusses illusions, colour, movement and the implications for the visual arts. A fascinating book.

HAMLIN, Sonya (1988): 'How to talk so people listen' (Harper & Row, New York, also Thorsons, Wellingborough, 1989)

An engaging book, written in a direct and informal style. It is highly structured with headings, making it easy to find what you want. It deals with a wide range of communication situations in business, including personal encounters, presentations, interviews and meetings. The emphasis is on planning each encounter by thinking about the role, motivation and personality of each participant. There is useful advice about communication on television. Throughout, there is good practical advice which encourages a sensitivity to the feelings of those involved. It relates to Chapter 3 here.

HANDY, Charles (1994): 'The empty raincoat' (Hutchinson, London, UK), also published as 'The Age of Paradox', Harvard Business School Press, Cambridge, MA.

Charles Handy is an acknowledged authority on the nature of work and business organisations. In this book he presents an engaging account of the trends in the 1990s and beyond. It relates closely to the discussion in Sections 1.2 and 7.4 here, about the future of business and the impact of IT systems on working practice.

'Technical writing for publication' (1992). Institution of Electrical Engineers, London

A detailed guide for authors of technical books or of papers for journals. It deals with the conventions of manuscript production, including typescript, spelling, illustrations, tables, photographs, mathematical notation, references, scientific units, copyright permissions, proof checking, indexing and production. It relates to Section 4.4 here. There is a companion pamphlet (see next item).

'Technical report writing' (1993). Institution of Electrical Engineers, London, UK
  An excellent short guide whose approach accords with Chapter 4 here. The pamphlet
  is based on ten 'laws' of report writing. The first is 'The reader is the most important
  person'. The last is the same; it is repeated because it is the only law which should
  never be broken. There is practical advice about format, references, writing,
  illustrations, tables, summaries and appearance. Although written in the context
  of technical reports, this pamphlet applies equally to other types of management
  report. There is a companion pamphlet (see previous item).

IND, Nicholas (1992): 'The corporate image: strategies for effective identity programmes'
  (Kogan Page, London, UK, 2nd edn.)
  Corporate public relations is an important topic for business today. It is peripheral
  to our main theme here, but the book is a good starting point for those who wish
  to explore this aspect further. It relates to Section 3.3 here.

JACKSON, Donald (1981): 'The story of writing' (Studio Vista, London, UK)
  A good representative of the many which deal with the historical development of
  writing, over 5000 years. It reviews writing techniques and the various styles which
  have given rise to those in use today. There is a good survey of contemporary
  penmanship, both for practical and artistic purposes. It develops the general theme
  of Section 2.2 here.

JANNER, Greville (1986): 'Janner on meetings' (Gower, Aldershot, UK)
JANNER, Greville (1988): 'Janner on communication' (Hutchinson, London, UK)
  Greville Janner is a lawyer, politician and businessman. He has written a wide range
  of books in the broad area of business communications. The details of two which
  are directly relevant here are given above; the first relates to Chapter 6 and the second
  to Chapter 2. The emphasis is on meetings and communications which involve the
  public (rather than on internal business communication). In those areas the detail
  complements that discussed here. For example the 1986 book deals with the rules
  of procedure of formal meetings, chairing public meetings, and conferences. There
  is useful material about visual aids, and the use of humour. The 1988 book is more
  closely related to the material here. However, there is useful additional information
  about public communication and the media. Dealing with the media requires special
  skills, and Janner offers good practical advice.

MCCALL, Ian, and COUSINS, John (1990): 'Communication problem solving: the language
  of effective management' (Wiley, Chichester, UK)
  Deals with an extension of the main topic: corporate communications within and
  between organisations. It relates human communication to issues of structure, politics
  and power base. It also deals with cross-cultural issues, relating to Section 3.3 here.

MEAD, Richard R. (1990): 'Cross-cultural management communication' (Wiley,
  Chichester, UK)
  This book addresses a topic which is vital for people working in multinational
  enterprises. Such people need cross-cultural skills when working within the enterprise
  itself, when negotiating with people outside the enterprise, at social occasions. This
  applies whether the individual is based at home or abroad. The book gives a systematic
  approach to communication in a wide range of business situations. There are diverse
  practical examples. The book is strongly recommended for those who are likely to
  find themselves in cross-cultural situations, or who need to understand the problems
  of others who face those problems. It relates to Section 3.3 here. See also Morris
  (1994).

MELLOR, D. H. (1990) (Ed.): 'Ways of communicating' (Cambridge University Press,
  Cambridge, UK)
  A collection of essays based on the 1989 Darwin College Lectures, on the subject
  of communication. There is a general introduction and eight main chapters, each
  by an authority in the field. The topics include the functioning of the brain, animal
  communication, language, literature, nonverbal communication, music, and the role
  of technology. A stimulating and wide-ranging book which enables the layman to
  draw links between different fields. It develops the ideas of Section 2.2 here.

MORRIS, Desmond (1994): 'The human animal: a personal view of the human species' (BBC Books, London, UK)

Desmond Morris is a zoologist who has become a popular author and presenter. This book supported a BBC TV series, and deals with a range of areas in which human behaviour is similar to, or has developed from, that of the animals. The first chapter of the book is called 'The language of the body', and deals with gesture. It is important to understand this when communicating across cultures. The book relates to Sections 2.2 and 3.3 here. For more detailed information, see Argyle (1988) and Mead (1990).

'Oxford dictionary of modern quotations' (1991): (Oxford University Press, Oxford, UK)

There are many dictionaries of quotations, and most people will have at least one on their bookshelves at home. For business use it is important not to overdo quotations; their use can be seen as ostentatious. However, an apt quotation can illuminate the communication process, whether spoken or written. You need a modern dictionary which is likely to include relevant business topics. This is a good example, with all the examples drawn from the 20th century. Some dictionaries are arranged by topic, with an index of authors; others use the reverse. With a good index it does not much matter which approach is used. See also the next entry.

'Pan/Chambers book of business quotations' (1987), compiled by Martin H. Manser (Pan Books, London, UK)

See also previous entry. This book is more business oriented. It is divided into nine main themed chapters, each subdivided into several topics. This is convenient for finding an appropriate quotation on a given subject. It is also good for browsing.

PEEL, Malcolm (1990): 'Improving your communication skills' (Kogan Page, London, UK)

This popular book has an approach similar to the one used here. There is less information about report writing and planning presentations, but a deeper discussion of face-to-face meetings (such as interviews), and dealing with the media (see Section 3.3 here). In that respect, see also Janner (1986 and 1988). Peel also discusses the use of communication technology. This has a wider agenda than Chapter 7 here but the discussion is generally briefer. Recommended as a complement to what is presented here. It is always good to have a different perspective on a subject like human communication!

POPPER, Sir Karl (1994/A): 'The myth of the framework' (Routledge, London, UK)

POPPER, Sir Karl (1994/B): 'Knowledge and the body-mind problem' (Routledge, London, UK)

POPPER, Sir Karl (1995): 'The open society and its enemies' (Routledge, London, UK)

Sir Karl Popper died in 1994. He was probably the greatest philospher of science this century. His masterpiece 'The logic of scientific discovery' was published in German in 1934, but not in English until 1959. 'The open society and its enemies' (1945, and reissued in 1995) was his argument against political authoritarianism. The first two titles listed were published after his death. They were edited by Mark Notturno, and cover Popper's later work as well. The connection here is with our brief discussion of logic in Section 3.4. Popper's view is that the information content of a theory is measured by what it denies, or rules out, rather than by what it asserts. This makes the theory falsifiable. If later evidence shows the theory to be false then we know we must change it. Evidence in support of an assertion, on the other hand, is never conclusive.

'Roget's thesaurus of English words and phrases' (1987): New edition prepared by Betty Kirkpatrick (Longman, Harlow, UK)

Roget's thesaurus was first published in 1852 and has sold over 30 million copies. This new edition deals with the unprecedented growth of English vocabulary caused by the scientific, cultural and social developments of the 1980s. The book has a system of classification which, like the Dewey subject classification, is of interest in its own right. However, most users will turn immediately to the Index, which runs to over 600 pages. This will lead directly to the meaning of the word itself, and to alternatives with slightly different meanings. The thesaurus will help you identify the word which conveys exactly what you intend.

SEEKINGS, David (1989): 'How to organize effective conferences and meetings' (Kogan Page, London, 4th edn.)

Much time is spent in meetings, not all of it usefully. This book deals mainly with large gatherings rather than small business meetings. It considers the preliminary planning, choosing the venue, selecting the programme and speakers, presentation techniques, supporting exhibitions, marketing, administration and running the event itself. There is a chapter about organising events overseas. The book is a useful guide for anyone responsible for arranging a large business or professional gathering. There are useful tables about room-sizes required for different numbers of people and seating arrangements. It relates to Sections 5.3 and 6.3 here.

SHANNON, Claude E., and WEAVER, Warren (1949): 'The mathematical theory of communication' (University of Illinois Press, Urbana, USA, reprinted 1963)

The classic text on information theory. It is a major text for communication engineers, and those interested in the mathematical approach (including the links between information and entropy). The book gives the key ideas of the process model of communication, and of the three levels of effectiveness of a communication process. It relates to Section 2.2 here.

SHEPHERD, Walter (1971): 'Shepherd's glossary of graphic signs and symbols' (Dent, London, UK)

A remarkable compilation, based on a simple system of classification, which accommodates more than 5000 individual symbols. The classification can be used to identify a given symbol; the index can be used to find a symbol with a particular meaning. Of interest in its own right, this book is useful to discover meanings (perhaps unexpected) which may attach to a proposed business logo. It relates to Section 2.2 here. For a more historical approach, see Whittick (1971).

SILK, David J. (1991): 'Planning IT: creating an information management strategy' (Butterworth–Heinemann, Oxford, UK)

Aimed at senior general managers, this book takes an overview of the task of information management in the business enterprise, using IT where appropriate. It develops the concept of management presented in Section 2.4 here, and shows how IT can give business benefit. It discusses six principles of information management; they can be viewed as a development of the issues discussed in Chapter 7 here.

SILK, David J. (1995): 'Harnessing technology to manage your international business' (McGraw-Hill, Maidenhead, UK)

Deals with how international businesses can use technology internally, or take advantage of the technological infrastructure. This covers information, manufacturing and logistics systems. There is background information about current technology and systems, as well as how to use them to best effect. It develops our discussion here in several ways: trends in modern business (Section 2.3); cross-cultural issues (Section 3.3); information systems, and how to use them to suppport group working (Chapter 7).

STANTON, Nicki (1986): 'Communication' (Macmillan, Basingstoke, UK, 2nd edn.)

The book deals with a wide range of business communication skills, and is directed particularly at those studying for professional qualifications. It is a detailed book, with over 400 pages and a wide range of exercises. You may find it helpful to explore further many of the topics presented here.

TURNER, J. Rodney (1993): 'The handbook of project-based management' (McGraw-Hill, Maidenhead, UK)

An excellent guide to the techniques of project management, from a general management perspective. It has been quoted in Section 3.3 here for its discussion of national cultural differences. There are some broad differences between developing and developed countries, but there are also more detailed differences which relate to individual countries and to the stages of project management involved. Turner's chapter on international projects is a good introduction to this topic.

WHITTICK, Arnold (1971): 'Symbols: signs and their meaning and uses in design' (2nd edn.)

A comprehensive historical survey of the development in the use of symbols. There are numerous examples, including many relevant to trade and business. The book relates to Section 2.2 here. See also Shepherd (1971).

WOOD, Alexander (1944): 'The physics of music' (Chapman and Hall, London, UK, 1975, 7th edn.)

The book describes the nature of sound and the mechanism of hearing. It deals with the musical scales and the functioning of musical instruments. Although it does not go far into the aesthetic dimension of music, it provides a valuable link between physical and musical principles. Engineers will be comfortable with the technical detail and enjoy the general approach. It develops a theme presented in Section 2.2 here

Index